A Basis for Scientific and Engineering Translation

A Basis for Scientific and Engineering Translation

German–English–German

Michael Hann
University of Mainz

John Benjamins Publishing Company
Amsterdam / Philadelphia

The paper used in this publication meets the minimum requirements of American National Standard for Information Sciences – Permanence of Paper for Printed Library Materials, ANSI Z39.48-1984.

Library of Congress Cataloging-in-Publication Data

Hann, Michael.
A basis for scientific and engineering translation : German-English-German / Michael Hann.
p. cm.
Includes bibliographical references and indexes.
1. Science--Translating. 2.Technology--Translating. 3.English language--Translating into German. 4.German language--Translating into English. I.Title

Q124.H36 H36 2004
501'.4-dc22 2003067672

ISBN 90 272 2608 3 (Eur.) / 1 58811 483 X (US) (Hb; alk. paper)
ISBN 90 272 2609 1 (Eur.) / 1 58811 484 8 (US) (Pb; alk. paper)

John Benjamins Publishing Co. · P.O. Box 36224 · 1020 ME Amsterdam · The Netherlands
John Benjamins North America · P.O. Box 27519 · Philadelphia PA 19118-0519 · USA

Table of contents

Acknowledgements

Special thanks are due to Jason Hann for excellent supplies of computer hardware and up-to-date software throughout the project, but above all for patiently converting an enormous collection of ragged Microsoft Word files into a neat, fully operational HTML electronic book. A debt of gratitude is also owed to Jean Hann for diligently proof-reading the entire manuscripts of both the handbook and the e-book. Grateful thanks are due to Bertie Kaal and friends, of John Benjamins Publishing Company, for permission to reuse valuable material from my earlier publication *The Key to Technical Translation* (1992). This book is dedicated to the memory of Stanley William Hann, who provided infinite support throughout his life.

Credit for many ideas employed in the lexicography units and dictionary arrangements is due to a group of writers discussed in the final Appendix. Some of these are personal friends and have provided valuable direct feedback or inspiration over many years: Juan Sager, Paul Kussmaul, Don Kiraly. Others belong to an earlier period, but their writings and ideas are universally familiar and well established: John Lyons, Eugene Nida, Eugen Wüster. Grateful thanks for helpful comments, direct and indirect inspirations relating to specific sections of the book are also due to certain colleagues and ex-colleagues at the Johannes-Gutenberg University of Mainz (FB23, FASK): Paul Foster, Rudolf Mikus, Eberhard Rüffer, Rainer Torka.

Inspiration and valuable feedback for the engineering sections came from several local scientific and industrial institutions, especially: the Nuclear Research Centre (KfK), Interunion-Technohandel, Watlow-Electric, AITEC. Technical translators need helpful customers and engineering partners with whom they can discuss their work at any time. Grateful thanks for assistance in this respect are due to Messrs Cook, Göbelbecker, Mikosch, Wiebe.

The most valuable literary inspirations, however, were obtained neither from academic nor industrial colleagues but from many hundreds of enthusiastic, highly gifted students taking the *technical translation* options in the language department (FASK) of the Johannes-Gutenberg University in Germersheim, Germany. Equipped with minimal resources, they struggle with a wide range of translation topics and regularly provide not only highly original translation solutions, but also valuable insight into the kind of literature and reference material urgently required by technical linguists. Grateful thanks are expressed for many constructive comments and suggested improvements on my earlier literary attempt in this field. Hopefully, this book will compensate enthusiastic translators at all levels by helping them to complete their education.

Preface

With the advent of on-line dictionaries, Internet searching facilities, e-mail and other electronic aids, the closing years of the millennium brought radical changes to many professions, but especially to that of the technical translator. Gone are the days when translators invested in huge personal libraries, printed dictionaries and well-guarded card indexes. The modern linguist has become a member of the new generation of "mouse-clickers", tapping computers for up-to-date first-hand information all over the world. This approach provides *rapid* translations, but a lot of time is spent *surfing* or merely *drifting around* on the Internet, and often the more unfamiliar the subject matter the less readable the ultimate translation becomes. Translators who use this method exclusively lack a systematic basis for their work, a knowledge of *technical language* (Ge. *Fachsprache*).

Learning a technical language, for instance the specialist languages of Mechanical, Chemical or Nuclear Engineering, is similar in many ways to acquiring the skills of communication for a foreign natural language. It is not just a question of *terminology*. Grammar rules change, verbs and prepositions acquire new significances, and similar terms occur with entirely different implications within brief sections of the same engineering text. Even a highly proficient translator with many years of experience in the field of Electronics, who can happily distinguish the fundamental meanings of the German polyseme *Widerstand* (E. *resistance, resistivity, resistor, reluctance, reactance, impedance*) within this specific field, has to be careful not to confuse the countable nouns (CN's) *resistance, reluctance, reactance, impedance* with their non-countable counterparts. This book and its electronic component examine not just the *engineering* basis of technical translation, but also the *linguistic* and *general semantic* aspects. They fulfil the functions of several books in one.

1. Layout

Modern translators are fully conversant with on-line methods of data exchange, whereby information requested in response to a query is accessed automatically and presented in little boxes. There are limits to the amount of information the user can truly absorb by this method, but it does have advantages when cross-referencing or re-accessing information. The *technical component* of the book, the chapters dealing with the conceptual aspects of engineering and the appropriate bilingual dictionaries, are confined to the disk, whereas the *linguistic component* appears in this handbook and provides a useful guide to the content of the disk.

There are three *disk volumes.* The first is mainly text subdivided into *chapters, units, sections,* and *subsections.* It also contains numerous *micro-glossaries* covering the individual engineering areas. The other two volumes consist of *large dictionaries* and *indexes.* A fourth area of the disk contains some 45 coloured *illustrations* clarifying important conceptual aspects of the engineering chapters. The discussion below distinguishes the trees from the ordered forest and examines the general structure of the book.

The first disk volume (Vol.1) contains 16 *chapters* headed:

1. Basic Mechanics
2. Basic Electricity
3. Materials Science
4. Nucleonics
5. Semiconductor Technology
6. Electronics
7. Circuit Technology
8. Automotive Engineering
9. Machine Technology
10. Chemical Engineering
11 Computer Engineering
12. Mechanics
13. Construction Engineering
14. Mechanical Engineering
15. Electrical Engineering
16. Mathematics

The chapters gradually describe the *basis* of the entire spectrum of scientific and engineering disciplines, providing fundamental terminologies of the areas concerned in both English and German.

Interspersed among these chapters is a second component consisting of eight so-called *lexicography units,* dealing with areas like:

1. Translational Equivalence
2. Interpretation
3. Dictionary Structures
4. Technical Grammar
5. Linguistics
6. Concept Analysis
7. Translation Difficulty
8. Specific Expression

These too are subdivided into sections and tackle other fundamental problems, encountered by technical translators, from a *linguistic* vantage point. These units have a variety of purposes, but one of their main objectives is to acquaint the reader at the earliest possible stage with the *dictionaries* of Volume 2.

The latter are not just dictionaries in the normal sense of the word but carefully organised term bases revealing the *polysemy* of technical language as well as *lexical, grammatical, contextual* and *semantic* information essential to translation proficiency. There are three major *dictionary units* headed:

1. Technical Polyseme Dictionary (TPD)
2. Technical Thesaurus (TT)
3. Technical Collocation Dictionary (TCD)

and two large *indexes*:

4. Main Index (MI)
5. German Index (GI)

which direct the *reader* (i.e. the *user* of the disk) to individual chapter sections containing terminology sought, via English (Main Index) and to a lesser extent via German (German Index).

The third volume has two main components:

6. German-English Alphabetic Dictionary (GE)
7. English-German Alphabetic Dictionary (EG)

each providing direct links to the dictionary resources of the other two volumes and the bilingual illustrations.

The handbook is divided into *units* discussing important aspects of the *disk chapters* and other features of the first two *disk volumes.* Some units, those dealing with *engineering,* constitute carefully watered-down versions of the *disk chapters* or combinations of these chapters. The rest, including the Introduction and the first Appendix, concentrate on material not present on the disk or less readily accessible in *disk sections,* aspects of *technical language* relating to areas like *grammar, lexicology* and *linguistic semantics* with occasional references to *text typology* and *translation theory.*

Obviously, continued usage of terms like *disk volume, disk section, disk subsection* would soon make the handbook appear repetitive. The expression *disk* is therefore dropped, and the disk itself is hereafter referred to simply as *the book.* The same convention is employed on the disk, where the terms *reader* and *user* become, in fact, contextually synonymous.

2. Objectives

The disk provides a summary of the underlying basis of science and engineering, of the associations and inter-relationships among technical terminology, and of the most persistent errors made by translators from or into German. It also provides concrete applications of *general linguistics* to technical literature, and of *structural lexicology* to the presentation of semantic information.

The book aims at the following main groups:

i. *professional translators* lacking a formal scientific or engineering education;
ii. *teachers* of technical translation strategies and expertise who seek additional inspiration for their courses from what could be regarded as a *teacher's* as well as a *translator's handbook*;
iii. *students* of technical translation or *language graduates* embarking upon a career in this field who wish to make a valuable long-term investment in their professional future.

The disk itself is a carefully structured multi-dimensional didactic tool, enabling readers to absorb vast quantities of technical and linguistic information passively at their own pace, permitting rapid access to previously absorbed material via a cross-referencing system originally geared to the printed page. A gradual transfer of information takes place from the book to the reader's brain, which copes with elaborate networks of conceptual links automatically and receives continual memory-refreshing boosts from the daily processes of reading, cross-referencing and general familiarisation.

Readers should use the disk initially as a self-teaching aid. Sections which take more time to digest can be printed and stored in a permanent hard-copy folder, where undisturbed by noisy fans and flickering monitor screens the reader can absorb the information efficiently. Acquiring the skills of technical translation with the aid of the disk is like learning to play a musical instrument. A direct link forms between the mind of the musician and the instrument played, between the reader and the book itself. This takes time, familiarity and practice. The handbook helps the reader to accomplish the first stage.

3. User Expectations

To the world at large a specialised book is a static entity, the contents of which correspond to the state of the art in the discipline concerned at the date of publication. But to an author it is a living organism evolving in tiny stages on disk and continually looking towards its next revision. One such book, entitled *The Key To Technical Translation* (Hann 1992) was published by the author during the first era of global PC activity.

In response to valuable feedback from many sources, the meticulous all-embracing project of reorganising and updating the original version commenced. This entailed expanding the spectrum of scientific and engineering fields covered, restructuring the dictionaries and incorporating far-reaching additions and improvements in regard to grammar, linguistics and lexicology. Fortunately, this period coincided with one of rapid advancement in data-processing technology itself and provided software presentation facilities which would have been almost unimaginable ten years beforehand. The long-awaited, ultimate result of the revision never came. Instead a new book emerged, one which in view of present approaches to translation activities was best presented as an *electronic book.*

In contrast to some e-books that are essentially glorified second-hand databases with information downloaded from different sources and *adapted* to the purpose at hand, the e-book is therefore the result of a gradual evolution of a very large, fully consistent *printed manuscript* designed to be read from end to end. To avoid unreasonable or false expectations, users accustomed to more conventional electronic information management environments, Web-based training materials, etc., who may be more interested at this stage in what the book *does not do*, are recommended to consult Appendix 3. Those prepared to study the information necessary to acquire the expertise for rapid efficient access during professional translation assignments will realise that the effort spent is well rewarded. Indeed, by and large, the easiest way to get the maximum benefit out of the book is to study the main sections *before* being confronted with urgent material for translation.

The need to acquire specific access expertise is not the only temporary obstacle impatient users will come up against, when using the disk. Obviously, the terminologies of English and German presented do not contain *every* isolated expression a technical translator might wish to find — a database of the entire terminologies alone of all branches of engineering, science and mathematics would comprise several million entries. Instead the disk resolves the most common misconceptions by covering in depth the set of several thousand *key terms* basic to all technical translation areas. It aims at a long-term improvement in the reader's overall translation proficiency by providing access to the likely missing link in virtually every current technical-translator education programme: the language of technology itself.

Introduction

There are hopeful translators who feel that electronic media will soon provide cost-free solutions to all their problems. But they are deluding themselves. Many hours of productive work can be wasted during on-line searching, and even then the results only consist of haphazard, fragmentary details. Under such conditions, "technical translation" remains little more than disappointing guesswork. No self-respecting translator would accept an assignment for a language he or she had no grounding in. By the same token, technical translators with a fluent command of general language but a poor command of technical language are living very dangerously.

Large-scale, multi-access, interactive term bases and other electronic data facilities provide intelligent suggestions for translations of *obscure* but well-defined compound technical expressions, but unfortunately they generally make little attempt to distinguish *basic engineering* conceptions which, by virtue of the polysemous nature of both technical and general language, can be even more difficult to translate precisely. For instance:

Auftriebskraft	buoyancy, lift, upward thrust
Kondensator	capacitor, condenser
Spannung	bias, emf, potential, voltage
Verkleidung	cladding, cowling, padding

Nonetheless, just as personal computers replaced the typewriters and card indexes of previous translator generations, so electronic media are clearly replacing libraries. This handbook and above all the accompanying disk therefore offer not a substitute for the use of data media but a parallel approach, one which provides a firm conceptual basis, constant didactic training and is likely to be more stimulating and rewarding intellectually than just slavishly scanning the Web all the time. It provides a systematic approach to the study of science and engineering from the translation aspect itself, applying features of the reader's likely educational background, areas such as *general linguistics*, *semantics*, *lexicology*, to the broad spectrum of technical information common to the overall majority of scientists, engineers and modern industrial technologists.

1. Broad Outline

Linguists with different areas of specialisation, scattered all over the world, easily communicate with one another in a language of their own, employing terms like *homonym*, *speech act*, *denotation*, *paradigm*, *phoneme*. The situation is similar in engineering, which employs terms like *energy*, *power*, *current*, *resistance*, *momentum*, with precise meanings understood by all technologists.

The next level of technical language requires a small degree of specialisation. A phoneticist employing expressions like *glottalic egressive*, *prominence peak*, *rhotacized vowel* and a grammar theorist studying *elliptic genitives* or *transitional conjuncts* may find initial communication slightly difficult, but it would not be impossible, as they share the same basic education. An electronics engineer too could easily discuss aspects of his job with a nuclear scientist, and vice versa. Moreover, if they happened to be involved in a joint project, each would rapidly acquire a knowledge of the other's terminology. A freelance technical translator must acquire this degree of familiarity with the terminologies of *all* engineering areas, and know instinctively and immediately which of a range of possible alternatives, such as:

Spur	lane, trace, track, wake
Stange	arm, bar, lever, rack, rod
Träger	carrier, girder, holder, member, rack

is the correct one in a given specialised context. The book aims specifically at these initial levels of technical language.

Beginning with the terminologies of Mechanics and Electricity, Volume 1 of the disk gradually covers the broad basis of all areas of science, engineering and mathematics employed in industrial technology. Volume 2 presents three highly concentrated, didactically organised glossaries for German/English/German translators that supplement this knowledge, each with an elaborate system of mnemonic descriptors providing intricate labyrinths of semantic information relating to *hyponymy*, *polysemy*, *synonymy*, *contrast*, *context*, *usage*, *association* and other features. The *lexicography units* take a look at other considerations involved in terminology specification and demonstrate applications of *linguistic semantics* (Lyons et al.) to the structural organisation of lexicographical information. The set of coloured illustrations containing terminology in two languages completes the reader's understanding of basic technical material.

This handbook provides a brief introductory description of the engineering chapters and summarises the ancillary sections. Certain material is not present anywhere on the disk, for instance Units 10, 14, 17, which present separate linguistic introductions to each of the main dictionaries, and Unit 19 that touches upon aspects of technical translation beyond *phrase-level.* Other units discuss material of general interest to technical linguists, especially to non-native English speakers.

The result is a contribution to the literature of scientific and engineering translation studies presented at two levels: a handbook providing a light introduction to German/English technical translation with a substantial general-linguistics orientation; an electronic book supplying an in-depth view of the scientific and engineering complexities but with a less pronounced linguistics component.

2. Engineering Chapters

The disk contains sixteen main chapters dealing primarily with *engineering* conceptions, but also with aspects of *general science* and *mathematics.* The first two chapters illustrate fundamental distinctions employed in the definition of engineering parameters, for instance *energy, force, impedance, power, voltage,* some of which occur in all branches of technology. These conceptions were formulated long ago by scientists, such as Newton, Ampere, Faraday, Maxwell and Kelvin, but still cause tremendous problems for inexperienced translators. Having mastered them, however, the reader can proceed to the next level of language specialisation.

Chapters 3–15 cover specific technologies, such as: Nuclear Technology, Electronic Circuit Design, Constructional Engineering, Aerospace Technology. Distinctions are drawn between *scientific* and *engineering* disciplines, such as Mechanics or Chemistry on the one hand and Mechanical and Chemical Engineering on the other. Attention is frequently focussed on problems resulting from the usage of identical expressions in different fields to refer to very different concepts (polysemy):

Fremdkörper	contaminant, impurity
Klemme	clip, clamp, terminal
Teilspannung	voltage component, stress component

Chapter 16 completes the reader's command of technical language by introducing certain important conceptions relating to all areas of science and technology but belonging to a quite separate self-contained academic field, namely *Mathematics.*

3. Terminology Sections

Within the engineering chapters there are subsections dealing with terminology, where important concepts like *average, mean, deviation, variance*, which are often confused by linguists with limited technical background, are defined explicitly. Concepts concerning concrete objects, *fuselage, hull, helm, rudder*, are described in context. Those difficult to visualise, such as *alkane, alkene, alkyne*, are described by analogy with everyday situations. Different translational possibilities or paradigmatic substitutions according to the level of language or potential customer are discussed (e.g. *Deponie*: landfill, disposal site, dump), and frequent translation errors occuring in each field are highlighted, especially where they result from similar terminology belonging to other fields:

Verbrennung	combustion, incineration
Widerstand	resistance, reluctance, reactance, drag
Zerfall	decomposition, disintegration, dissociation

These sections stimulate the reader's enthusiasm for the dictionaries of Volume 2, where other potential pitfalls for translators are revealed:

Basisstärke	basicity
Beleuchtungsstärke	illumination intensity
Feldstärke	field strength
Stromstärke	current

In normal engineering contexts, substitutions like *"base strength", *"illumination strength", *"current strength" would be quite wrong.

As well as dealing with *contrast* among terminology (e.g. *emf/mmf, fatigue/creep, isomer/polymer*), there are subsections within the engineering chapters illustrating *grammatical* and *lexical* aspects too. For instance: Chapters 1 and 2 illustrate semantic differences between *countable* nouns such as *mass*, *charge*, *speed*, *light* and their corresponding *non-countable* but orthographically identical counterparts; one aspect of Chapter 8 concerns *polyonymy* in the Automobile Industry, a phenomenom whereby identical car parts sold on opposite sides of the Atlantic have different names; Chapter 11 draws attention to the *lives*, *recent deaths* and *altered significances* of specific terminology in fields such as Electronics or Computer Engineering which change very rapidly.

The organisational structure of the engineering chapters is designed to permit smooth transitions from the explanation of scientific and technological conceptions to the discussion of translation problems and the avoidance of translation errors. To facilitate equally smooth transitions for access purposes during subsequent reading, readers are recommended to re-examine the *Contents* section of the disk at periodic intervals and print a hard copy.

4. Lexicography Units

The lexicography units (Lex. 1–8) placed strategically among the engineering chapters take a closer look at fundamental terminology and direct attention to important structural features of the dictionaries of Volume 2. These units complete the reader's mental digestion of scientific or engineering information by examining *grammatical*, *lexical* or other *linguistic aspects* of terminology covered up that point. They discuss *translational equivalence*, the semantic connotations of *noun countability/non-countability*, the use of collocations to resolve *syntagmatic* problems in translation and the application of conceptually organised terminological hierarchies to overcome *paradigmatic* ones. The units examine linguistic aspects of terminology that are useful for concept specification and differentiation, such as *synonymy*, *hyponymy*, *polysemy*, *homonymy*, *context*, *usage*, and take a brief look at the type of writing styles and conventions to which most technical translations conform.

As the reader's command of specialised concepts and understanding of technology progresses, more advanced applications of semantics are introduced in order to discuss unusual *technical linguistic* phenomena, such as different *knowledge structures* in different countries in the same engineering area. Other sections analyse different *degrees* of translation difficulty and the gravity of specific substitution errors. The Appendix takes a small break from Engineering and illustrates applications of the collocational and thesaurus approaches to dictionary organisation for other areas of specialised translation, such as Business Studies.

5. Dictionary Units

The second volume is dominated by the three elegantly structured dictionaries (TPD, TT, TCD). These employ a wide variety of symbols, designating the following features: fields, subfields and areas of science or engineering; semantic relationships among terminology (e.g. *hyponymy*, *partitive association*); different categories of noun (CN, NCN, ...). The dictionaries themselves are discussed in due course, and so is the set of dictionary symbols, three tables headed *Field Codes*, *Thesaurus Descriptors*, *Other Symbols*. A clear understanding of the semantic principles governing these symbols helps the reader to interpret information more efficiently in the actual dictionaries.

Another feature of Volume 2 is the index component. Facilities for accessing terminology, whether English or German, *directly* in the engineering chapters or lexicography units appear in the *Main* and *German Indexes* respectively. The book is arranged so that any important *engineering terminology* appearing in the handbook reappears in a similar context on the disk, thus averting the necessity of additional handbook indexes.

Volume 3 contains two more dictionaries, this time conventional alphabetic ones — German-English (GE) and English-German (EG). Hasty readers used to working with simple data bases might mistakenly assume that these dictionaries are the ultimate goal of the book. They are in fact just a by-product, presenting the terminology of the other dictionaries, thesauri, indexes, diagrams and illustrations *collectively* in a more readily accessible form. Instantaneous access to the sources themselves is achievable via the eye buttons.

6. Lexicology Units

One question invariably asked by students struggling to learn two or more foreign languages while being confronted simultaneously with the classic work of authors like Lyons, Leech and Nida is "*Why do we have to learn so much linguistics?*". The answer is "*Because it is useful*". And this book proves it.

Using *natural language* to illustrate different degrees of *hyponymy*, distinctions between *antonymy* and *contrast*, or different categories of *synonym* has the disadvantage that *terms* are not as tightly bound to *concepts* as in technical contexts. The *lexicography units* of the disk are not subject to this drawback. They familiarise the reader with *general semantic* aspects of new terminology, stimulate the reader's enthusiasm to use the dictionaries, and encourage individual translators not only to interpret structural clues regarding *concept specification* in terminological data bases, such as the TPD, but also to think and act like a lexicographer.

Some aspects of these lexicography units are discussed again in this handbook. So too are certain features of the dictionaries TPD, TT, TCD. Indeed, the handbook contains units that effectively constitute *new* dictionary introductions, from a slightly more advanced linguistic vantage point. Units 10 and 14 concern the first two dictionaries and adopt a classical semantic approach, discussing: *homonymy* as opposed to *polysemy; hierarchic* versus *non-hierarchic* conceptual relationships; *morphological* considerations relating to term selection. Unit 17, the second TCD introduction concentrates more on *concordances* concerning prepositions and nouns, nouns and verbs, on phenomena involving *technical verbs* and *specialised predicates*, and on the *narrow significances* of adjectives and adverbs in specialised contexts.

Hence some sections of the handbook deal not with specific background information relating to technical translation but with general lexicology. In many respects, these *lexicology units* are as important to the dictionary user as the dictionaries themselves.

7. Long-Term Objectives

For too long, academic institutions have treated technical language as jargon separate from natural language. It is not. Technical language evolves naturally. It simply entails conceptual restrictions unfamiliar to many linguists. The book traces its evolution from natural science to the basis of modern industrial technology and reveals how labels attached to concepts in one language differ radically from those in another. The prime objective of the two volumes is to provide an efficient, systematic method of learning the skills of technical translation, a viable alternative to the blind substitution employed by too many on-line dictionary enthusiasts and translators with no formal training in engineering whatever.

A second, long-term aim of the disk publication is to produce a book whose organisational structure could function as a permanent model for general reference, in regard to:

i. future literature on technical translation (Engineering, Science, Mathematics);
ii. similar material on other areas of specialised translation (Law, Medicine, Business, Economics);
iii. material for language pairs other than English-German.

The book is suitable for all technical translators from student to professional, regardless of background or ability, by virtue of the manner in which information contained is absorbable in various ways. Despite the electronic appearance readers are advised to work systematically through the book, studying one chapter at a time, and obtain familiarity with the organisational structures of the dictionaries at the earliest possible stage. But at their own pace, as the structures themselves are of valuable assistance in the gradual acquisition of technical language skills.

The book has a final primary objective. It focusses attention on the lexical basis of engineering in the hope that gifted translators will eventually be in a position to obtain foreign-language equivalents of obscure compound terms, not included on the disk, by *inspired* guesswork. Substitutions may need to be verified by scanning international data-base networks, but, in contrast to the usual, haphazard, hit-or-miss methods of Internet surfing, the reader will then have a better idea of whether the term selected is *likely* to be correct.

8. Technical German Text Samples (TGTS)

Further sources are available on the author's web site, including a collection of German source texts for initial classroom practice for the training of technical translators. This material is useful for an elementary understanding of technology, thereby employing terminology fundamental to the areas concerned. The texts vary in length from single paragraphs to 500 words, sample a broad range of science and technology.

User Guide

Many readers will wish to study the handbook first and then work their way systematically through the e-book, familiarising themselves with the access techniques as they do so, in which case they should skip this section and proceed to the first Unit. Those who wish to acquire broad familiarity with the e-book immediately should follow the guidelines below.

Insert the disk, go to your CD-Rom drive and click on e-start to get to the Home Page.
The CD is PC and Mac compatible. For optimal viewing please adjust your Internet Explorer settings to View → text size → smallest.

1. Textual & Glossary Material

A click on *Volume 1* reveals access to the *engineering chapters* (Chap. 1–16) and *lexicography units* (Lex. 1–8). Click open a chapter; the first *page* appears. Chapter sections are opened by links at the bottom of the first page, the *chapter page*, subsections at the bottom of the respective *section page.* To return to a section from a subsection click on the *diagonal arrow* near the top left corner of the window. To return to the chapter page from a section, or to the chapter list (*Volume 1* page) from the chapter page, do likewise. Some pages are long and require scrolling. At the foot of a long page, click on the *vertical arrow* to return to the top. All *diagrams, tables, microglossaries* and *microthesauri* relevant to the chapter are accessed via links at the foot of the *chapter page.* The basic procedure is the same for the *lexicography units, dictionaries* and other units of the two volumes.

The *navigation bar* at the top of the page enables the user to switch to other parts of the book (*Volume 2, Illustrations,* etc.) at any time, or to consult the *Dictionary Symbols.* Illustrations themselves are opened by clicking on their respective *thumbnails.* The *dictionaries* and *indexes* are divided into pages according to the initial letter of the entry term like a conventional dictionary.

They are opened at the page required by clicking on the book icon in the *vertical bar* (left). The user can scroll to the entry sought or skip to it directly via the letter buttons of the *horizontal bar* at the top of the page, returning to the top at any time via one of the many *vertical arrows.*

It is advisable to consult the *Contents* of the e-book at regular intervals. The Contents are accessible via the main navigation bar and provide direct access to any section of the book. The user should also get accustomed to the *eye buttons* (*blue, purple, green, brown*) that provide alternative access to each of the four main areas of the disk:

Volume One	16 engineering chapters, 8 lexicography units
Volume Two	3 large dictionaries, 2 indexes
Volume Three	2 large reference dictionaries
Illustrations	45 technical diagrams and illustrations

The reader should also study the main introduction, the separate introductions to Volumes 2 and 3 and the various dictionary introductions in depth. The *red eye* (Hints) draws attention to important sections that require more frequent scrutiny: *User Instructions, Search Instructions,* etc.

2. Browsing

It is possible to open the e-book at different points simultaneously, in different windows. Thus *subsections* of the same or different chapters can be compared, *chapter sections* can be browsed alongside *diagrams, term lists, microglossaries,* or alongside the main *dictionaries* and relevant *illustrations.* The normal browser commands, i.e. *Back, Find, Home,* etc., apply throughout the book.

Find (CTRL-F) is especially useful. Supposing the reader is looking for possible English translations of the German technical term *Läufer.* One method is to open up the Technical Polysyme Dictionary (TPD) at page *L* and then click on *La* in the horizontal bar. If the desired equivalent is not obtained, try the German Index. Another way is to use *Find* and enter *"läuf"*. The German Index reveals the information:

Läufer 8.2.1 Figure 8B Figure 8F

Using the same input data, still present in the *Find* command, the reader discovers English equivalents at various locations elsewhere on the disk: in context in Chapter 8, in the structured term list Figure 8B and in the micro-thesaurus Figure 8F. A similar approach for another expression:

Laufwerk 11.5 Figure 11H eins-

reveals, in addition, a location on page *E* of the Technical Collocation Dictionary (TCD) (entry *einschalten*). Readers accustomed to research involving *web pages* will soon discover many similar shortcuts for smooth rapid access using the cross-referencing facilities provided.

3. Visual Material: Illustrations

Readers are now recommended to familiarise themselves with the coloured *illustrations* present on the disk. They are divided into the following groups:

Basic Mechanics
Basic Electrical Engineering
Materials Science, Nucleonics, Semiconductors
Electronics, Electronic Circuit Design
Automobile Engineering
Chemical Engineering

but there are many overlaps. The illustrations link terminology in both English and German to visual diagrams. Readers struggling to grasp moderately difficult engineering concepts — *momentum, reaction, phasor, RMS voltage, reactance, lattice bond, power supply, diode characteristic, rocker shaft, alternator, alkane* — who are familiar with these diagrams, may find them a great asset when they have trouble digesting information of the e-book or its handbook counterpart.

4. Access, Navigation

Volume 1 can be read, page by page, like a conventional book by clicking on either the subsection links or navigation arrows according to the instructions given. Alternatively, direct access to a particular section is obtained by clicking on *Contents* in the navigation bar and selecting the appropriate line in the Table of Contents. A third method, providing instantaneous access to any chapter, section or subsection from anywhere on the disk is via the *eye buttons* (Chap. 1–8: *green*; Chap. 9–16: *brown*). This navigation technique is recommended when using the indexes. (See Appendix.)

The dictionaries of Volumes 2 and 3 are divided into *pages* according to the initial *letter* of the entries. They are opened in a similar manner (*letter buttons*, *navigation arrows*, etc.). A second method facilitating rapid access between any of the seven dictionary units is to employ the *blue eye* button. Instantaneous access from *any page* of a dictionary to any page of *another* dictionary is via the *purple eye*.

The blue eye lists the dictionaries in the order of presentation:
TPD TT TCD MI GI GE EG

The purple eye separates the letter pages into two blocks:
TPD TCD GI GE TT MI EG

according to whether the dictionary entries are German or English.

The blue eye also provides direct access to the full set of *microglossaries* of Volume 1 and to all *illustrations*. Indeed, together, the four eye buttons provide immediate access from any electronic page of the entire disk (*chapter*, *illustration*, *thesaurus*, *index*, etc.) to any other. They are complemented by another set of buttons that reveals the various *dictionary symbols* (field codes, thesaurus descriptors, etc.) used throughout.

The disk volumes vary in size. Together with the *illustrations*, Volume 1 is responsible for about two-thirds of the information contained, Volume 2 about one-third. Volume 3 merely repeats terminology present elsewhere on the disk and provides direct access to more detailed information in the other volumes.

The disk is user-friendly but unconventional.

Brief reminders of techniques explained in the introductory sections are obtainable via the *red-eye* button (hints).

5. Systematic Study

Before shutting down the computer and returning to the handbook, readers should click on the icon of the main navigation bar to return to the *home page*, and then click open the main *Introduction*. There are three subsections — *Primary Objectives*, *General Layout*, *User Instructions*, each of which should be studied properly, possibly by printing a hard copy. They illustrate the structure of the disk and demonstrate the full range of electronic tools provided.

Translators might not study their Bible every day, but should get into the habit of gradually or periodically working through the disk, *page by page*, in consecutive order as if it were truly a book. It makes translation under pressure so much easier. This handbook provides a more leisurely introduction to the main features of the e-book, to general problems encountered in technical translation and to slightly advanced usage of the disk from a linguistic viewpoint.

Note

Though possibly itself a useful contribution to the world library of technical translation literature, this handbook serves a primary purpose: the description of the e-book on the CD-Rom that accompanies it. To avoid confusion, the reader should bear in mind at all times when hastily re-examining individual passages that, unless otherwise specified, the terms *book, disk, chapter, section, subsection* refer only to the e-book. Cross-referencing within the handbook itself is minimal but, where necessary, takes place via the terms: *handbook, handbook unit, unit, handbook section.* Likewise the terms *illustrations* and *figures* refer to different sections of the e-book, the expression *diagram* being used only in reference to their content.

Appendix: Symbols and Abbreviations

Navigation Symbols

	Homepage
	Field Codes
	Thesaurus Descriptors
	Other Symbols
	Unit Symbols
green	Chapter Sections 1–8
brown	Chapter Sections 9–16
blue	Figures, Illustrations, Dictionary Homepages
purple	Individual Dictionary Pages
red	Navigation Hints, American Terms

Dictionary Symbols

Table 1. Field Codes

ACU	Acoustics
AERO	Aeronautical Engineering
ASTR	Astronomy, Cosmology
ATOM	Atomic & Molecular Physics
AUTO	Automobile Technology
BATT	Batteries, Battery Cells
BIKE	Bicycles
BRAK	Automobile Braking Systems
CHEM	Chemistry, Chemical Engineering
CLOK	Clocks & Watches
CONS	Construction Engineering
DPS	Data Processing Systems
EENT	Electronic Entertainment
ELEC	Electricity, Electrical Engineering
ELNC	Electronics
ELSC	Electrostatics
EMAT	Engineering Materials
ENGN	Engines
FLUD	Fluid Mechanics/Fluid Dynamics
FRIG	Refrigerators & Freezers
FUEL	Automobile Fuel Systems
GAS	Gases & Vapours
GEN	General Language
GEOM	Geometry
GRAF	Graphs & Charts
HOUS	Household Applications
HYD	Hydraulic Engineering
IGN	Automobile Ignition Systems
LAB	Laboratory Apparatus
LAMP	Lamps & Fittings
MACH	Machine Technology
MAGN	Magnetism
MATH	Mathematics
MATS	Materials Science
MEAS	Precision Measurement
MECH	Mechanics, Mechanical Engineering
MET	Meters & Gauges
METR	Meteorology
NAUT	Nautical Engineering, Shipbuilding
NUCL	Nucleonics, Nuclear Engineering
OFF	Office Equipment

OPT	Optics
OSCN	Oscillations & Vibrations
PHOT	Photography, Camera Systems
PHYS	Physics
POW	Power Systems
RADN	Radiation, Radioactivity
RAIL	Railway/Railroad Engineering
REM	Remote Control Systems
ROCK	Rocket & Missile Technology
RUNN	Automobile Running Gear
SDEV	Semiconductor Components/Devices
SEMI	Semiconductor Design Technology
STEA	Steam Engines
SUBJ	Subject Field, Academic Discipline
SUBS	Substances, Materials
TELE	Telecommunications
TOOL	Tools, Implements
TRAN	Transformers
TV	Televisions, Monitors
WAST	Chemical/Radiochemical Waste Disposal
WAVE	Waves, Wave Propagation

Table 2. Thesaurus Descriptors

Symbol	Implication	Example
a:	*associated with*	anode (ELEC), a: battery cell.
ct:	*contrasted with…*	atomic number (CHEM), ct: mass number, valency.
cv:	*covers the concept(s)*	electromagnetic wave (PHYS), cv: radio wave, light wave.
co:	*consist(s) of*	electron cloud (MATS), co: freely mobile electrons.
d:	*defined as/designates*	disintegration (NUCL), d: rapid nuclear decay.
ex:	*typical example…*	electrical quantity (PHYS), ex: current, voltage.
m:	*measurable parameter of*	electrode gap (AUTO), m: spark plug.
p:	*part of…*	element (ELEC), p: heater, cooker, kettle.
s:	*synonym/abbreviation*	HT-lead (AUTO), s: ignition lead.
t:	*a type of …*	magnistor (ELNC), t: magnetic sensor.
tu:	*typical unit …*	pressure (PHYS), tu: bar, millibar, pascal.
u:	*used in connection with*	protractor (MATH), u: angle measurement.

cs: contextual synonym/near synonym
ps: preferred term/preferred synonym
nps: non-preferred term

Table 3. Other Abbreviations

sg	Singular Noun: *physics, statics, wave mechanics*
pl	Plural Noun: *electrics, nuclear binding forces*
CN	Countable Noun: *gearbox, gradient, integral*
PN	Pair Noun: *callipers, dividers, pliers, shears*
NCN	Non-Countable Noun: *ammonia, chromework, wiring*
CN/NCN	Dual Noun: *charge, current, energy, power*
Br.	British: *alternator (AUTO), truck (RAIL)*
Am.	American: *AC generator (AUTO) , wagon (RAIL)*
Br./Am.	Both Variants: *disk (DPS), generator (AUTO)*
obs.	Obsolete Expression: *atomic weight, condenser*
lmn.	Layman Expression: *breadboard, engine revs*

Table 4. Unit Symbols

amp	A	ohm	Ω
milliamp	mA	kilohm	kΩ
microamp	μA	megohm	MΩ
volt	V	micron	μ
millivolt	mV	angstrom	Å
microvolt	μV		
kilovolt	kV	coulomb	C
megavolt	MV	electron-volt	eV
watt	W	newton	N
milliwatt	mW	newton-metre	Nm
megawatt	MW	joule	J
farad	F	henry	H
microfarad	μF	weber	Wb

Unit 1

Access Facilities

At this point readers divide naturally into two broad classes: those impatient to click open sections of the electronic dictionaries to see what *terminological* information the book supplies, and those preferring a more systematic approach aimed at acquiring an overall long-term improvement in their translation proficiency by virtue of the *conceptual* information provided. This unit is for the dictionary enthusiasts. It serves as an initial guide to the cross-referencing and organisational facilities of the disk, discussing collective features of the *microglossaries* following the early engineering chapters and the various large *dictionaries*. Discussion of dictionary search strategies and other ancillary work involved in technical translation continues at various points later in the handbook, employing examples from the engineering chapters discussed up to that point.

1.1 Technical Polyseme Dictionary

The Technical Polyseme Dictionary (TPD) is a structured bilingual glossary designed to reveal the polysemous nature of German technical terminology. Entries are arranged alphabetically in order of *root terms*, followed by their respective *compounds*. The objective is to make the following attributes of technical terminology explicit:

polysemy	*Netz* (network, grid, graticule, reticle)
	Lack (paint, lacquer, varnish, photoresist)
homonymy	*Bahn* (lane, trajectory, webbing)
	Scheibe (pulley, washer, window pane)
hyponymy	*Kraft* (force), *Antriebskraft* (propulsion),
	Auftriebskraft (buoyancy), *Gewichtskraft* (weight)
	Säge (saw), *Ansatzsäge* (tenon saw),
	Laubsäge (fretsaw), *Stichsäge* (compass saw)

Other attributes appear distinctly too:

contextual synonymy	*Methan-/Ameisensäure* (formic acid)
	Photodissoziation/Photolyse (photolysis)
antonymy/contrast	*Plus-/Minuspol* (positive/negative terminal)
	Eigen-/Störstellenleitfähigkeit (intrinsic/extrinsic conductivity)

Polysemy presents the main difficulty in translation assignments, and is the feature most obvious in this dictionary, by virtue of the different L2 equivalents. Hence, the name: Polyseme Dictionary.

1.2 Technical Thesaurus

Each engineering chapter is followed by one or more microglossaries or microthesauri summarising the terminology of the chapter. The large Technical Thesaurus (TT) supplements the smaller thesauri and introduces other engineering conceptions not necessarily based on Volume 1.

For instance, the expression *carriage* has the engineering interpretations:

i. part of a typewriter or printer
ii. part of a lathe
iii. part of a railway train

Concepts of this type are distinguished in the Thesaurus by field codes, DPS, MECH, RAIL and other *specific information.* In order to save space, however, and concentrate the reader's attention on the engineering associations, as well as other *entry links* in the thesaurus, *printer, lathe, railway,* frequently employed phrases of the type "part of" are replaced by symbols such as *p.*

The dictionary user rapidly learns new symbols by heart (e.g. *t:* "a type of"), and is able to apply them to less familiar concepts in the Thesaurus:

auger	t: hand tool for boring holes.
autoclave	t: purification apparatus using superheated steam.
NTC resistor	t: device whose resistance decreases with temperature.

The symbols themselves are listed in Table 2 of the *Dictionary Symbols.*

1.3 Technical Collocation Dictionary

There are verbs whose significance is restricted to a single engineering interpretation, such as *leerlaufen* (to idle), a term whose meaning is associated only with automobile engines. But most verbs encountered by technical translators reappear in a variety of a different contexts. These too have precise interpretations and require a specific translation from a range of possibilities:

ausdehnen	extend, expand
lösen	solve, resolve, dissolve
verstärken	amplify, magnify, strengthen

Adverbs can also be troublesome, so can adjectives, prepositions, and even certain general nouns when their meaning is narrowed by a specific engineering context.

Whereas the other dictionaries highlight *technical polysemes*, the Technical Collocation Dictionary (TCD) draws attention to the polysemous nature of *general vocabulary* in technical contexts. Entries provide *examples* for resolving translation problems in specific contextual environments.

1.4 Noun Classes

Non-native English speakers beginning to learn the skills of general translation into the foreign language tend to produce statements like:

*"The scissors is broken."
*"The police has arrived."
*"The goods is outside in the van."
*"The data requested were faxed yesterday."
*"The informations are in this book."

all of which contain *serious errors*. They illustrate rare cases where grammatical considerations outweigh semantic ones; the sentences should read:

"The scissors *are* broken."	– implying *one pair* of scissors
"The police *have* arrived."	– *a single* helpless policewoman
"The goods *are* outside in the van."	– *one box* containing a camera
"The data requested *was* faxed yesterday."	– the *data figures* requested
"The information *is* in this book."	– the *information details* requested

German has *der/die/das* to make life difficult for non-native speakers, but English too has several important noun classes which can only be mastered by obtaining familiarity with the language and studying the grammar. Indeed, the above examples themselves illustrate a variety of noun classes described by grammarians by labels such as: *singular/plural, countable/non-countable, collective* and occasional alternative designations such as *uncountable noun, mass noun.*

Some words commonly regarded as non-countable nouns, like *energy, impedance, mobility,* often switch categories in an engineering context and behave as countable nouns:

> A single accelerated particle acquires *an energy* of 30 eV.
> The speakers have different *impedances.*
> Electron *mobilities* vary according to the material structure.

Expressions like *work* (Ge. 1. *Arbeit;* 2. *Werkstück(e)*) do not:

> *Some work* is done on the gas, possibly as much as 28 kJ.
> *The work* done by the engine per revolution amounts to 500 J.
> *Certain work* is treated for rust protection even before machining.

The dictionaries indicate this aspect of technical language. Nouns which are always *singular* or always *plural* are followed by the symbols *sg/pl* respectively. They are also differentiated as follows:

CN	Countable Noun: *carburettor, network, reaction, winding*
NCN	Non-Countable Noun: *bonding, chromework, damp, flux*
PN	Pair Noun: *callipers, pliers, dividers, shears, tongs*

The symbols are used in all dictionaries and thesauri throughout the book, so that the non-native English speaker realises immediately that expressions like **chromeworks,* **paintworks,* **bondings* do not exist in standard technical English whereas *networks* is perfectly acceptable. Their inclusion also helps the dictionary user to separate extended meanings of similar terms, a feature especially useful in analysing fundamental technical expressions, such as *charge, inductance, resistance, tension,* which function as NCN's when referring to an *engineering property* and CN's when denoting a measurable *parameter.*

1.5 Microglossaries, Thesauri

Diagrams, such as the Periodic Table of Elements or the Electromagnetic Spectrum appear at the ends of the chapters concerned. They are followed by alphabetic or hierarchic lists of terminology, in both English and German, or by thesaurus arrangements illustrating terminological relationships. Various sorting configurations are employed.

Consider, for example, the following arrangement:

1	radiation	*Strahlung*
11	thermal radiation	*Wärmestrahlung*
12	electromagnetic radiation	*elektromagnetische Strahlung*
13	nuclear radiation	*Kernstrahlung*

Even though the reader may not be familiar with the concepts at this stage, it is evident at a glance that the concept *radiation* is divisible into three broad categories: *thermal, electromagnetic* and *nuclear radiation.* A slight addition at the appropriate point in the arrangement:

121	microwave radiation	*Mikrowellenstrahlung*
122	optical radiation	*Lichtstrahlung*

reveals the next level of subdivision: *electromagnetic radiation* relates to either the *microwave* or *optical* part of the spectrum; *optical radiation* is itself divisible into *infra-red, visible* and *ultra-violet radiation.* A simple alphabetic glossary of *radiation* terms would occupy the same amount of page space, but not reveal conceptual interrelationships. Where convenient, therefore, the book employs *hierarchic* arrangements.

Taking two further examples, this time without the German:

1	travelling wave	1	atom
11m	amplitude	11p	set of electron shells
12m	wavelength	111p	electron(s)
13m	propagation velocity	12p	nucleus
14m	frequency	121p	proton(s)
		122p	neutron(s)

In the first example, *amplitude, wavelength,* etc. are not hyponyms of *travelling wave* but *measurable parameters* (m) describing the wave. Similarly, an *atom* can be considered to consist of the following *constituent parts* (p): a set of *electron shells* and a *nucleus,* themselves containing one or more of the constituents *electron, proton, neutron.*

A slightly different method of illustrating similar information is the thesaurus approach. Information relating to associated concepts appears in concise form alongside entries:

amplitude	m: travelling wave
nucleus	p: atom
proton	p: nucleus
visible radiation	t: optical radiation

Most chapters are concluded by detailed *microthesauri* similar to the above. In addition, some contain *glossaries* of hierarchically interrelated terminology.

1.6 Classification Systems

The *hierarchic* approach is merely an alternative application of the common *family tree* structure used by genealogists to reveal lineages of ancestral heritage. These methods are also employed by biologists to indicate taxonomic relationships among plants, creatures and organisms, by library workers classifying and arranging books, and above all by *information scientists* and *software writers* dealing with data-management schemes requiring fast retrieval and compact storage.

Some linguists would no doubt prefer the usual *tree diagrams* employed by semanticists to illustrate what they call "*lexical*" or "*semantic fields*": the set of *concepts* denoted by a given superordinate lexeme (*dog*: *bulldog, poodle, terrier,* ...) or the set of *attributes* (e.g. *dog*: *paw, tail, faithful, smelly*), where set elements are further divisible into subsets. But for engineering applications, and especially for the more sophisticated conceptual hierarchies appearing in the later chapters, the enormous space requirements together with the increased complexity and amount of repetition required rule out tree diagrams themselves. The standard method of indicating hierarchical classifications:

1.2.2.2.3.g omega particle

is also avoided, in the book, in favour of:

12223g omega particle

in order to provide room for near synonyms and lengthy German equivalents on the same line, since many readers will presumably wish to print out individual arrangements and study them at their leisure.

It is helpful for readers if they aquaint themselves with hierarchic organisation at an early stage. The same system of notation is employed later in the book in connection with more advanced semantic discussion relating to the *structural organisation* of the dictionaries, as well as for illustrating different *knowledge structures* in different parts of the English and German-speaking worlds. Numerals indicate which pair of terms is related; four symbols (a, g, m, p) reveal the kind of relationship — *associative*, *generic*, *metric*, *partitive*, and correspond to the definitions: "associated with", "type of", "measurable parameter of", "part of".

1.7 Main Index, German Index

The Main Index (MI) specifies locations of English terms within the engineering chapters, rather like the *index* of a normal specialised book. It has entries like:

tension 1.4.4 2.4 8.1 13.1

which reveals that the term concerned appears in Chapter 1 (Subsection 1.4.4), Chapters 2, 8 and 13. The index itself has many other functions, providing access to terminology in the *lexicography units*, the *collocation dictionary* (*TCD*), as well as in various *term lists*, *microglossaries* and *thesauri*. The latter generally provide possible German equivalents together with structured definitions. Where TCD collocations exist, they lead the translator indirectly to German equivalents in context.

The German Index (GI) provides direct access to TCD collocations of German terminology too. But its main purpose is to direct the reader to *microglossaries* and *thesauri* containing this terminology, and to other specific sections of engineering chapters where it occurs. As the German Index contains fewer entries than its counterpart, some entries reveal a certain extravagance. For example:

Dehnung: Figure 13 Figure 13 Figure 13 Figure 13

This type of entry is not a misprint. It means that the term concerned appears at least four times in the same microglossary. If the glossary is a *thesaurus* the reader may find four separate definitions or descriptions of the different concepts involved (e.g. *elongation*, *extension*, *strain*, *tensile strain*).

Thus, like the other dictionaries, both the MI and the GI illustrate the *polysemy* of technical language. This is a fortunate by-product of their structural organisation but a benefit translators should not overlook.

1.8 Alphabetic Dictionaries

Systematic translators consult more than one dictionary and, when searching for L2 equivalents, they attempt to specify the exact *concept* implied by using whatever reference sources are available. Regrettably, another species of translator appeared at the turn of the millennium, one who clicks L2 terms into translations from a trusted electronic dictionary almost without bothering to read them, the *non-thinking* translator. Ironically, both are well-served by the two *alphabetic dictionaries* of Volume 3, German-English (GE) and English-German (EG).

The dictionaries are intended for the more systematic translator and are merely logical extensions of the first two volumes providing direct reference to terminology in the TPD, the TT, the various microglossaries, microthesauri, bilingual diagrams and illustrations. Entries are specified by their *field* and more precisely by their *source* (TPD, Figure 2A, etc.) in separate columns of the dictionary table. Like the indexes (MI, GI), the alphabetic dictionaries (GE, EG) provide indirect indications of terms which are *polysemous*. It is up to the user to follow the leads provided.

Unit 2

Basic Mechanics

Chapter 1 of the disk divides into sections according to the following scheme:

1.1 Statics, Kinematics, Dynamics
1.2 Physical Quantities
  1.2.1 Basic/Derived Quantities
  1.2.2 Mass, Weight
  1.2.3 Energy, Power, Work
  1.2.4 Coulomb, Kelvin, Candela, Mol
  1.2.5 Unit Conventions
  1.2.6 Orthographic Conventions
1.3 Scalar/Vector Quantities
  1.3.1 Magnitude, Direction
1.4 Mechanical Quantities
  1.4.1 Speed, Velocity, Acceleration
  1.4.2 Power, Performance, Efficiency
  1.4.3 Impulse, Momentum
  1.4.4 Stress, Strain, Tension
  1.4.5 Moment, Torque, Torsion
1.5 Units, Symbols
1.6 Grammar

Figure 1: Mechanical Quantities

The chapter begins by introducing some fundamental scientific terminology from the field of *Physics*, and from an important subfield: *Mechanics*.

2.1 Physics, Mechanics

The study of Physics constitutes an important part of the basic education of all scientists and technologists, and the terminology of this discipline appears throughout Engineering. It contains many branches. Whereas terms relating to Optics or Acoustics, for instance, are needed only by specific translators working on material involving cameras, loudspeakers or musical instruments, the terminologies of two major physics branches, *Mechanics* and *Electricity*, are fundamental to virtually all technology, and underlie Mechanical Engineering, Electrical Engineering, Semiconductor Electronics, Aeronautics, Automotive Technology and many other industrial areas. Chapter 1 focusses on *Mechanics*, one of the oldest technical disciplines of all, which dates back to Aristotle's *law of levers* and his deductions of *buoyancy*, and now provides technology with the means to explore the beds of the oceans and send people to the moon. The second chapter concentrates on *Electricity*.

2.2 Multiple Meaning, False Friends

Because physics terms are used in so many different engineering areas, they have adapted over the course of time to these areas. Polysemous terms which cause confusion among technologists are eventually modified, but the changes implemented are by no means symmetrical from one language to another. Scientists living a century ago were no doubt content with translators who rendered *elektrische Spannung* as *electrical tension*. Modern technologists, however, faced with the obsolete terminology of their great grandfathers, rapidly become frustrated and irritated on reading a text which obliges them to make the repeated mental substitution *voltage*. Moreover, renderings like **electrical voltage*, **mechanical tension* as translations of *elektrische/mechanische Spannung* can leave a similar bad impression on the translation customer. He knows of no other kind of *voltage*, no other form of *tension* springs to mind, and the result may be chaos anyway if the translator is unaware of the significance of the second technical meaning of *mechanische Spannung*, namely *stress*.

There are many everyday terms which have a precise significance in elementary physics but are not included or given misleading translations in dictionaries. A conscientious translator given a complex technical text to translate may render the obscure terms correctly, yet his translation may be totally misleading because he does not realise that an apparently simple word like *Spannung* implies both *tension* and *stress* within the same short text. The alternatives denote very different concepts. And there are "false friends" too. The German expression *Impuls* is not translated by *impulse*, but by *pulse* in Electronics and *momentum* in Mechanical Engineering. The chapter investigates other German expressions, which have a variety of interpretations with different English equivalents, terms like *Drehmoment, Geschwindigkeit, Leistung,* together with common expressions whose meanings are often narrowed within science and technology by concise mathematical definitions: *Arbeit, Dehnung, Energie, Wirkungsgrad* (E. *work, strain, energy, efficiency*).

2.3 Parameter Definition

The opening sections of the chapter illustrate certain universal conventions employed by scientists and engineers in the definition of *technical parameters,* viewed from the aspect of their basic conceptual and mathematical properties. The expression *parameter* is actually an engineering concept referring to a subset of what scientists call *physical quantities* (Ge. *physikalische Größe*). The chapter concentrates on *mechanical quantities,* those central to the Physics branch Mechanics and to Mechanical Engineering, but introduces *basic quantities* relating to other branches as well (e.g. *charge, temperature, luminous intensity*) for the purposes of comparison. Translators equipped with the underlying principles of parameter definition can employ this knowledge to distinguish different concepts which happen to correspond to the same German expression, and to select the appropriate L2 equivalent in cases where *units* or other *physical properties* provide the only clue. For example: *tension, stress* (Ge. *Spannung*); *speed, velocity* (Ge. *Geschwindigkeit*); *power, performance* (Ge. *Leistung*). The bilingual (English-German) microglossary denoted by Figure 1, at the end of the chapter, summarises these clues for all terminology discussed up to that point. Figure 2A provides a similar service distinguishing the *electrical quantities* of Chapter 2, such as *voltage/potential, resistance/resistivity, reactance/impedance.*

2.4 Parameter Differentiation

Many technical translators are baffled by the enormous variety of *parameters* occuring in engineering texts. These are often crucial to the meaning of the target text and hence to the customer's understanding. The chapter offers guidelines which, though not applicable in every case, will nevertheless help translators reach speedy decisions regarding accurate L2 terminology.

For instance, suppose that a translator is unsure how to translate the expression *Bremskraft* in a text dealing with the testing of a new automobile engine. Dictionaries reveal three alternatives: *braking force, braking performance, braking power*. The translator looks closely at the source text and observes that the parameter concerned is measured in kW. After reading Chapter 1, he is aware that *force* is normally measured in *newtons, power* in *watts*, and that *performance* is not a well-defined engineering parameter at all. He selects *braking power*, which is likely to be the correct substitution.

The fact that no correlation exists between the units *newton* and *kilowatt*, and hence that there is a large difference between *braking force* and *braking power* may be obvious, even to the most humble laboratory assistant or automobile mechanic, but is a vital clue sadly missed by many translators. To rectify this problem, the chapter takes a close look at how parameters are defined in general, and how they can be differentiated. Distinctions are drawn between *scalar* and *vector* parameters, such as *speed* versus *velocity*. And the chapter relates any parameters sharing the same units to their common root concept — thus *tension, traction, thrust* designate different types of *force*.

2.5 Concept Determination

Even at this early stage many readers will feel overwhelmed by the complexity of technical language. But there is worse to come. It may take weeks to study the opening chapters properly, but the reader then has a valuable asset for the rest of the book and has made an indispensible long-term investment in a future career in translation. Chapter 1 illustrates the first examples of *polysemy*, drawing careful distinctions between the technical terms, such as *stress, strain, tension* or *moment, torque, torsion*. It examines fundamental conceptions like *mass/weight, speed/velocity, impulse/momentum*, which provides a sound basis for the differentiation of more complex engineering concepts elsewhere on the disk, for instance *atomic mass/atomic weight, orbital speed/orbital velocity*.

It is assumed from the outset that the reader has no real grasp of Mathematics. Hence, in order to provide a basis for differentiation among terminology, certain *conceptual tools* employed by scientists and engineers in everyday situations are also described in passing: *basic/derived quantities, scalar/vector quantities, magnitude/direction*. This enables the chapter to indicate how divisions occur naturally within Mechanics itself, areas such as *Statics, Dynamics, Kinematics*, and how these divisions lead on to the major engineering branches discussed in subsequent chapters: *Machine Technology, Construction Engineering, Transport Engineering* and of course *Mechanical Engineering* itself.

The reader soon realises that the immense variety of *units* employed in industrial technology (i.e. *joule, farad, millirem, megavolt, gallon, decibel,* etc.) virtually all derive from a tiny set of just five *basic* units, *kilogram, meter, second, coulomb, kelvin,* characterising the conceptions: *mass, distance, time, electricity, temperature.* The first three lead to other fundamental conceptions, *force, velocity, energy, power, work,* whose quantisation is independent of the context of the conception and applies equally to areas unconnected with the mechanical sciences, for instance *magnetic energy, electrostatic force, luminous power.* By a fortunate coincidence, these simplifications gradually thrust upon scientists and engineers over the years, in reaction to the complexities of their subject areas, provide useful guidelines for translators. Instead of merely *guessing* which of a range of dictionary alternatives is appropriate in a practical translation situation:

Gewichtskraft	weight, gravity, force of gravity
Geschwindigkeit	speed, velocity, rate
Impuls	impulse, momentum, pulse
Leistung	power, performance, output, capacity, efficiency

linguists familiar with the chapter acquire a *feeling* for these conceptions and make sensible intuitive translational substitutions from semantic clues provided by the conceptions themselves.

2.6 Units, Symbols, Orthography

Although different systems of units are used in different parts of the world, technologists are usually fully conversant with the various differences between British and American units (gallon, pound, etc.) and can easily convert units from one system to another. To avoid confusion, the chapter advises the reader *not* to change units in translated texts. The same applies to algebraic symbols, for instance *voltage* in English texts usually has the symbol *V* as opposed to *U* in German. The translator may wish to add a footnote, but the symbols themselves represent only a mathematical abstraction and should *not* be altered.

Some linguists would disagree with this advice and insist that the target text be fully adapted to the customer's vernacular working environment. But, in contrast to their engineer customers, technical translators rarely have a head for figures. Nor is their command of mathematics generally of a level where symbol substitutions can be carried out automatically. A miscalculation involving units can mislead and irritate the customer, just as much as a poor translation.

Failure to realise that a changed symbol (*V*) occurs in the source text with a different significance (e.g. *Geschwindigkeit*) could make nonsense of equations and cause complete confusion. In specific areas, such as Automobile Advertising, it may be helpful to change for instance *fuel consumption* in *litre/100km* into *mpg* (miles per gallon). But care is necessary. Otherwise the translator might unwittingly describe a *speedboat* with the fuel consumption of a *chain saw*.

The more specialised a unit becomes, for instance *becquerel*, *candela*, *weber*, the more likely it is that technologists will employ the singular form, especially in the spoken language. It is also the case when the individual concerned is subconsciously thinking of the *unit symbol* as opposed to the unit itself. This contrasts with the situation in general English which always quotes units in the plural: "a man six *feet* tall", "a tank with a capacity of 50 *litres*", "two light bulbs providing 100 *watts*". German is consistent but always quotes units in the *singular*: "drei *Pfund*", "fünfzig *Ampere*".

There are no hard and fast rules for Technical English as to which type of substitution is preferable. Translators must adopt a common sense approach. This is a minor problem but the chapter nevertheless provides guidelines to eliminate it.

2.7 Grammatical Distinction

Having discussed at length a variety of important conceptions fundamental to all branches of science and engineering, the chapter closes on a lighter note and approaches an area which many translators barely consider: that slight discrepancies occur in technical language in connection with *grammar*.

Consider the following *general* statements:

1. The vehicle loses *speed* on a sharp incline.
2. The vehicle is moving at *a speed* of almost 70 mph.

In the first example, the term *speed* is used as a *non-countable noun* (*NCN*). Substitutions like *some speed*, *a lot of speed*, *part of its speed* are perfectly acceptable but *"the vehicles lose speeds" is not a sensible, well-formed English statement. Example 2 illustrates *speed* as a *countable noun* (*CN*). Here the substitution possibilities are completely reversed.

Similar considerations apply to technical terminology. For instance, the expression *mass*:

3. In contrast to photons, electrons have *mass.* (NCN)
4. The electron has *a mass* of 0.9×10^{-30} kg. (CN)

Some of the terminology introduced in the chapter has a similar dual function. The NCN denotes a *technical property* of the specified object, and the CN a *parameter* specifying this property:

Electrons have *mass.* (NCN)	… *a mass of* 0.0005 a.m.u. (CN)
The particles acquire *energy.* (NCN)	… *kinetic energies of* 5eV. (CN)
The wire has *resistance.* (NCN)	… *a resistance of* 0.1 ohm. (CN)

Nevertheless, it is not advisable for the reader to seek sweeping generalisations. Similar statements can refer to the same conception:

The *mass* of the comet is rapidly decreasing. (CN)
The comet is rapidly losing *mass.* (NCN)

or have very different connotations:

Opposite *charge* attracts. (NCN)
Opposite *charges* attract. (CN)

The NCN *charge* denotes an *amount* of electrostatic charge present for instance on a cloth, metal rod, etc. The CN *charges* refers to discrete *quanta* of electric charge carried by atomic and other elementary particles (Chap. 3), a term used figuratively, as if "particles of *charge*" existed as separate, concrete entities.

It is recommendable for the reader to examine *verbs* that co-occur with NCN alternatives. Statements like:

*"electrons have charge"
*"the strut has tension"
*"the object has velocity"

do not make sense. Other substitutions are necessary or intelligent paraphrasing is required:

Electrons *carry* charge./… are *charged particles.*
The strut *is under* tension./… *is subjected to* tension.
The object *is moving* (at a particular velocity).

In isolated cases, attention to semantic aspects may also be necessary:

The filament generates *heat.* (NCN)
The liquid has absorbed *a heat of* 50 joules during this time. (CN)
The heating element provides *a continuous heat of* 2 kW. (CN)

The concept *heat* (NCN) is specified by two parameters denoted by two different CN's *heat*: one relating to *energy* and employing the unit *joule*, the other concerning *power* measured in the SI unit *watt*. The examples illustrate the typical laboratory report jargon of engineers and physicists analysing *heat transfer* or *heat flow* in a specific engineering system. Substitution of alternative expressions employing just the NCN *heat*, such as *2kW of heat*, *50J of heat*, enables the semantic problem to be sidestepped but leads to excessive repetition in translations.

Physical quantities can be divided into four groups according to the grammatical behaviour of the terminology concerned. Some terms are used only as CN's, for instance *angle*, *area*, *height*. Some, such as *inertia*, *gravity*, *work* are always NCN's. Thus statements like: *"a gravity of 50t", *"a work of 50kJ" are completely impossible and must be paraphrased:

a force of 50t due to gravity
50kJ of work

Terms like *momentum*, *pressure*, *stress*, *tension*, *thrust* appear throughout Engineering as both CN's and NCN's, with closely related meanings comparable to those quoted above for *mass*, *charge*, *heat*. Other CN/NCN terms, such as:

light NCN (optical energy)	Ge. *Licht, Lichtenergie*
light CN (a lamp, torch, etc.)	Ge. *Lampe*
sound NCN (acoustic energy)	Ge. *Schall, Schallenergie*
sound CN (a sound, note, etc.)	Ge. *Ton, Geräusch, Laut*

have different interpretations and different German equivalents.

The purpose of the *chapters* is to describe engineering concepts themselves, as simply as possible and without too much sidetracking. Grammatical guidelines for technical translators are therefore restricted to the *lexicography units*. Specific information is also available in Volume 2, where the Polyseme Dictionary (TPD) and Thesaurus make clear distinctions between CN's and NCN's, and the Collocation Dictionary (TCD) illustrates and contrasts different *usages* of fundamental engineering terminology.

By now the reader will have realised that there is another reason for referring to the sections of this handbook as *units* rather than *chapters*. They are in no way complete, but merely provide a means of light relaxation for the long dark evenings when the computer is off and the disk is safely tucked away. Before moving on to the next unit, load the disk again, examine Chapter 1 in greater depth and then click open some of the sections of Chapter 2.

Unit 3

Basic Electricity

Chapter 1 of the disk enables readers to understand what is meant by the engineering conception *parameter*. It introduces a small set of concepts which recur throughout the entire spectrum of Engineering, such as *energy*, *power*, *work*, *velocity*, *acceleration*, *momentum*, and fundamental *units* from which all other units throughout science and technology are derived: *candela*, *coulomb*, *kelvin*, *newton*, *joule*, *watt*, etc. The chapter is entitled Basic Mechanics, although it also includes Basic Physics, areas like *Optics*, *Acoustics*, *Heat Flow*, which have less to do with Mechanical Engineering itself. Nevertheless, one important group of fundamental engineering parameters is virtually absent, the *electrical parameters*.

Chapter 2 completes the picture, introducing the reader to the electrical sciences and the domain of Electrical Engineering. The structure is as follows:

2.1 Current, Voltage, Charge
  2.1.1 Voltage, Potential, Emf
  2.1.2 Bias, Tension
  2.1.3 Charge, Load
2.2 Resistivity, Resistance, Resistor
  2.2.1 Variable Resistor, Potentiometer, Rheostat
2.3 Direct/Alternating Current (AC/DC)
  2.3.1 AC Source, DC Supply
  2.3.2 Terminology Conventions
2.4 Reactive Devices
  2.4.1 Capacitor, Capacitance, Capacity
  2.4.2 Inductor, Inductance
  2.4.3 Transformer, Choke, Coil, Winding
2.5 Power, Wattage, Rating
  2.5.1 Peak/RMS Amplitude
  2.5.2 Real/Reactive Power

Some section headings may plunge readers who had difficulty with Chapter 1 into a state of panic, but remember that the book can be read on different levels. Translators specialising in Semiconductor Technology or Electronics need to be familiar only with basic distinctions, for instance *capacity/capacitance*, *resistivity/resistance*, *inductance/inductor* and the different interpretations of the highly polysemous German expression *Widerstand* (i.e. *resistance*, *reactance*, *impedance*). Those dealing with other electrical fields, *motors*, *generators*, *power transmission*, etc., need to be conversant with all of the chapter, but after clicking open and perusing the sections, readers can study them in detail at a later stage.

3.1 Subject History

The scientific study of *electricity* began in the early nineteenth century, largely in response to Volta's experiments with *electric cells*, but also inspired by a general growing awareness among physicists of the fundamental associations of electricity with *magnetism* and with what is now known as *static charge*. Michael Faraday and other great pioneers soon showed that the three new, rapidly emerging, scientific fields of *Electrochemistry*, *Electromagnetism* and *Electrostatics* did not involve three kinds of electricity (chemical, magnetic and static) but only one. When Faraday's demonstrations of the relationships between electricity and magnetism had acquired general recognition throughout the scientific world, and his analysis of the features and properties of electromagnetism became widely acknowledged, brilliant mathematicians such as Maxwell began their quest for concise, elegant *vector equations* to link the basic conceptions of electricity with the behaviour of *electromagnetic waves.* By the end of the century, Edison's invention of the *electric light bulb* had provided incentives for the construction of *power generators* and *power distribution networks,* Marconi was about to demonstrate the applications of *radio waves* to the field of "wireless transmission", and the way was clear for other far-sighted engineers with a keen eye on consumer markets to investigate the many practical advantages of electricity which eventually brought a vast range of household appliances, from hair dryers to satellite receivers, into the modern domestic environment.

Despite the long history of this field, certain basic concepts, such as the differences between *current* and *voltage, conductance* and *conductivity, charge* and *load,* are poorly understood by many translators. These fundamental electrical conceptions underlie numerous branches of technology and reappear throughout the book. They are the main topic of the chapter and reappear in the bilingual microglossary headed Figure 2A.

3.2 Voltage, Potential, Emf, Bias, Tension

Chapter 1 of the disk (Basic Mechanics) introduces the German polyseme *Spannung,* which is translated by either *stress* or *tension* according to the mechanical context. When the term refers to a *parameter,* the choice is easily resolved by close attention to the *units* employed (Figure 1). The same polyseme occurs in electrical contexts (i.e. *elektrische Spannung*), for which many conventional dictionaries simply present a vague list of alternatives: *bias, emf, potential, tension, voltage.* Here a different approach is required to differentiate them:

i. ***Voltage, Potential***

The expression *potential* occurs in Chapter 1 in connection with so-called *potential energy,* for instance the energy contained by a compressed spring (Ge. *Sprungfeder*). The chapter begins by drawing an analogy between objects rolling down a hill, losing the *potential energy* they initially possessed by being at the top of the hill, and an electrical situation involving *current* and *voltage.* A 9-volt battery creates a *potential difference* of 9 joule/coulomb (similar to a hill 9 metres high) across any component connected directly to the two terminals. The term *potential* appears in this field as a near synonym of *voltage* but with important restrictions. The English technical expression *potential* designates *relative voltage* with respect to a specific reference level in the electrical system or circuit, the so-called *earth* (Am. *ground*), which may or may not correspond to the *mains earth* (Ge. *Erdpotential*). Thus an electronics engineer who probes *potentials* at different points in a complex integrated circuit is conducting an activity rather like an aviation pilot recording the *altitudes* of different mountains from an aircraft and registering them with respect to a given level, for instance *sea-level.* It is evident from the analogy that the term *potential* necessarily implies *direct voltage* (Ge. *Gleichspannung*) in a *direct current* system; in connection with *alternating voltages* the term is meaningless.

ii. ***Voltage, Emf***

Another near synonym of *voltage*, used in connection with batteries and sometimes magnetic devices, is *emf* (also *e.m.f.*). The term originally stood for *electro-motive force* and implies the *voltage* generated by a particular *chemical* or *magnetic* arrangement. When a voltage is applied to a coil of wire wound around an iron core the flow of current is *impeded*, in the tiny fraction of a second required to set up the magnetic field, by a reverse voltage due to the magnetic arrangement itself. This is known as the *back emf* (Ge. *Gegenspannung*). A similar effect occurs when the magnetic field collapses. The *back emf* can easily reach values as high as 30 kV, for example the voltage produced by the *ignition coil* of an automobile, that causes the *spark plugs* to fire.

A different meaning of *emf* occurs in the field of Electro-Chemistry, where a specific arrangement consisting of an *electrolyte* (normally a certain *acid*) and two *electrodes* of different metals can constitute an *electric cell* — like the cells of a *battery*, each generating an *emf* of about 1.5 volts. Despite its rather nebulous definition and apparent double meaning, English-speaking technologists are reluctant to discard the term *emf* in favour of simply *voltage*, possibly because it provides a convenient stylistical alternative when formulating complex reports.

iii. ***Voltage, Bias, Tension***

Electronics reports and other documents compiled by English-speaking engineers contain expressions like: "a *bias* of 0.72 volts is sufficient to switch the diode into the conducting mode"; "the circuit requires a *bias* of at least 30V". The term *bias* is used here merely as a short form of *bias voltage.* Chapters 6–7 discuss the usage of *bias* as a contextual synonym of *voltage*, along with areas of overlap and contrast concerning the related expression *operating voltage.* In German, *Betriebsspannung* denotes both *bias* and *operating voltage.*

There is yet another term for *voltage*, which dates back to the pioneer days of electricity and still survives in the layman language of a few isolated areas outside Electrical and Electronic Engineering. The word is "tension". One example is the area Automobile Ignition which employs terms like *HT-circuit, LT-lead, LT-connection* (HT/LT = high/low tension). But here too, *voltage* is the preferred term and indeed the only possibility in terms like: *charging voltage, spark voltage, battery voltmeter.*

3.3 Entity, Property, Parameter

Broadly speaking, the translator may render *Spannung* as *voltage* in any electrical context. Translators who are aware of the existence of terms like *back emf, primary emf, battery emf* can employ them by all means, but semantic confusion will not necessarily result if the more obvious alternatives *back voltage, primary voltage, battery voltage* are substituted.The terms *bias, emf* and *potential* are pragmatic alternatives in specific cases, but *tension* is obsolete and should be avoided at all costs.

The variety of possible substitutions for the German concept *elektrische Spannung* does not present a serious problem to translators. The situation is very different, however, when another closely related polyseme is considered: *Widerstand* (Eng. *resistor, resistivity, resistance, reactance, reluctance, impedance*). Here the meanings are entirely different, and the result of a translation where substitutions like these are picked at random is chaos.

i. ***resistor, resistance, resistivity***

The first two alternatives *resistor/resistance* are easily distinguished. They conform to a conceptual pattern followed by other devices (e.g. *capacitor, inductor*) but to varied extents. The pattern is that of *entity, property, parameter.* A *resistor* is an object (i.e. an *entity*) which can be picked up and handled. *Resistance* (NCN) is a *property* of this entity, and *resistance* (CN) is a *parameter* characterising the entity. *Resistivity* (NCN) is a property of the *material* constituting the entity and *resistivity* (CN), a parameter characterising the material property. The pattern is similar for *capacitor/capacitance* and for *inductor/inductance.* However, the device properties *capacitance/inductance* have no simple corresponding material property. There is no concept *"capacitivity" or *"inductivity", analogous to *resistivity,* despite the frequent occurrence of the latter expression in many published dictionaries as a mis-translation of the German *Induktivität* (E. *inductance*).

ii. ***reactance, impedance, reluctance***

Reactance is employed only for specific electrical devices, *capacitors, inductors, chokes,* whereas *resistance* is reserved for devices whose behaviour in response to an *alternating voltage* obeys the same physical laws as for *direct voltage* (Ohms Law, etc.). The term *impedance* is the parameter employed when a device exhibits a combination of the properties *resistance, reactance.* But this is a radical oversimplification. Chapter 2 sheds more light on the matter. It is helpful to realise, nonetheless, that unlike resistance, reactance depends not just on the parameters characterising the electrical device concerned, i.e. *capacitance, inductance,* it also depends on the *frequency* of the alternating voltage bias. These concepts occur repeatedly in the book, and this discussion is continued on the disk. *Reluctance* is a concept related to *inductance,* but from the field of Electromagnetism rather than Electricity, and is not measured in the unit *ohm.* It is dealt with in Chapter 15.

3.4 Noun Countability

Let us consider at this stage an extract from the microglossary Figure 1, reproduced from the book, without the German equivalents:

The glossary enables a number of common translation problems to be resolved simultaneously. For example:

i. Two possible substitutions for the German polyseme *Spannung,* namely *stress, tension* are easily differentiated. The two parameters have different *units* (i.e. *newton.m*$^{-2}$ as opposed to *newton*).
ii. False friends are easily detected too, such as *Impuls* (E. *momentum*), *Kraftstoß* (E. *impulse*), even though the units may be the same.
iii. Closely related but non-synonymous concepts, such as *speed, velocity* (both *Geschwindigkeit* in German), are differentiated by the general parameter distinction *scalar/vector* introduced in Chapter 1.

What is interesting to the linguist is that some terms are *always* NCNs, for example *friction, gravity, work,* while others exhibit dual properties (CN/NCN): *energy, mass, momentum, power, pressure.*

Figure 2A provides a similar bilingual microglossary for the *electrical* parameters of Chapter 2. It too enables polysemes to be differentiated by their *units,* and by a similar parameter distinction *scalar/phasor* analogous to that of the mechanical quantities, for instance: *resistance/resistivity, conductance/*

Figure 1. *Extract*

Quantity	v/s	SI-Unit
acceleration	v	$m.s^{-2}$
angle	s	radian
angular velocity	v	$radian.s^{-1}$
area	s	m^2
compression	v	newton
compressive stress	v	$newton.m^{-2}$
density	s	$kg.m^{-3}$
distance	s	m
energy CN/NCN	s	joule
force	v	newton
friction NCN	v	newton
gravity NCN	v	newton
impulse	s	$kg.m.s^{-1}$
mass CN/NCN	s	kg
moment	v	newton.m
moment of inertia	s	$newton.m^2$
momentum CN/NCN	v	$kg.m.s^{-1}$
orbital velocity	v	$radian.s^{-1}$
power CN/NCN	s	watt
pressure CN/NCN	s	pascal
reaction	v	newton
rotational speed	v	rev/s,rpm
speed	s	$m.s^{-1}$
strain	s	DL
stress	s	$newton.m^{-2}$
tension	v	newton
tensile stress	v	$newton.m^{-2}$
thrust	v	newton
torque	v	newton.m
torsional stress	v	$newton.m^{-2}$
traction NCN	v	newton
upward thrust	v	newton
velocity	v	$m.s^{-1}$
volume	s	m^3
weight	v	newton
work NCN	s	joule

v/s = vector/scalar; m = metre; s = second; DL = dimensionless
The above terms are CN's unless otherwise specified.

conductivity, inductance/inductive reactance. Why then does it not reveal which terms are NCNs. The answer is that a more subtle differentiation technique is necessary, revealed later in the book (Lex.4–6).

3.5 Terminological Relationships

Whereas, in Chapter 1, it is possible to use examples involving simple technical expressions like *crane, elevator, bridge, steering wheel* which do not detract from other engineering concepts constituting the main theme (e.g. *power, performance, strain, torsion*), Chapter 2 requires the reader to visualise concepts like *impedance, inductance, capacitance, resistivity* on the basis of examples involving devices such as the *transformer, rheostat* or *choke,* which are not common everyday objects to non-technical readers. Though there are a variety of useful circuit illustrations, additional photos and diagrams would be difficult to accommodate on the disk and would not greatly contribute to a better understanding, since the reader's main interest as a translator lies in the terminology itself. The solution offered is to employ a more sophisticated type of glossary (Figures 2B–D), one which lists these additional basic concepts not alphabetically but in the form of semantically organised hierarchies.

Consider, by way of example, the following list which appears in the glossary of Figure 2C:

4	transformer	*Transformator*
41p	winding	*Wicklung; Spule*
411g	primary winding	*Primärwicklung*
412g	secondary winding	*Sekundärwicklung*
414m	number of turns	*Windungszahl*
42m	turns ratio	*Windungsverhältnis*

The hierarchic arrangement indicates that the semantic relationship between *winding* and *transformer* is a *partitive* one (p): a winding is "part of" a transformer. Similarly the relation between *primary winding* and *winding* is a *generic* one (g): the primary winding is *one type of* transformer winding; the other type is *secondary winding.* The expressions *number of turns* and *turns ratio* denote *metric* concepts (m), in other words measurable parameters normally involving *units*; in this case they happen to be *dimensionless quantities.* It is intuitively apparent in the hierarchy that the concept *number of turns* is equally applicable to the subordinate generic concepts of *winding,* i.e. to *primary winding* as well as to *secondary winding,* whereas the expression *turns ratio* (a parameter relating to both primary and secondary windings simultaneously) refers only to *transformer* itself.

Close examination of the second hierarchy of Figure 2B reveals other interesting features:

1	alternating voltage	*Wechselspannung*
11m	frequency	*Frequenz*
111m	angular frequency	*Kreisfrequenz*
12m	amplitude	*Größe*
121m	peak value	*Scheitelwert*
122m	RMS value	*Effektivwert*
1221a	RMS voltage	*effektive Spannung*
13m	phase	*Phase*
131m	phase angle	*Phasenwinkel*
1311a	voltage phasor	*Spannungszeiger*
13111a	phasor diagram	*Zeigerdiagramm*
2	alternating current	*Wechselstrom*
222m	RMS current	*effektive Stromstärke*
2311a	current phasor	*Stromzeiger*

For instance, that the concept *alternating voltage* is characterised by three metric parameters *frequency, amplitude, phase,* which are themselves characterised by other metric concepts, such as *angular frequency, RMS value, phase angle.* The hierarchy also reveals that the concept *alternating current* is closely *associated* (a) with *alternating voltage,* and if the reader were to guess from the presence of *RMS current* and *current phasor* that terms like *frequency, angular frequency, peak value, phasor diagram* can also refer to alternating current, the guess would be right.

It has taken a page or more to expand information contained in just a few lines of a hierarchic glossary, even without mentioning the German. Yet this is by no means irrelevant. The hierarchies of Figure 2C and 2D provide clear indications of the *polysemous nature* of the German terms *Widerstand* and *Spule,* even within this small basic field. They also illustrate why the English concept *resistance* requires four or more German equivalents, i.e. *ohmscher Widerstand, Verlustwiderstand, Wirkwiderstand, Gleichstromwiderstand,* and why *impedance* needs at least two: *Wechselstromwiderstand, Scheinwiderstand.* Once these hierarchies are intuitively expanded, properly digested and analysed, the reader can deduce other discrete packages of fundamental electrical knowledge directly and avoid many persistent conceptual errors and terminological inadequacies in future translation assignments.

Unit 4

Translational Equivalence

At this point it is useful to briefly revise some of the terminology of the first two chapters and aquaint the reader with the Technical Thesaurus and Technical Polyseme Dictionary (TPD) of Volume Two, in order to establish strategies for the access of terminological and lexicographical information which are applicable to other fundamental engineering concepts dealt with in subsequent chapters. This section is very similar to the disk unit *Lexicography 1*, and the terminology is accessible on disk in much the same form. It is reproduced here so that the reader can compare different lexicological approaches, and digest them, without too much mouse clicking.

4.1 One-To-One Equivalence

Terms such as the following from Chapter 1 are unlikely to create problems for translators:

acceleration	*Beschleunigung*
angle	*Winkel*
density	*Dichte*
energy	*Energie*
inertia	*Trägheit*
mass	*Masse*
pressure	*Druck*
strain	*Dehnung*
volume	*Volumen*
weight	*Gewicht*
work	*Arbeit*

For each English term there is one German equivalent, and vice versa. The odd term might acquire an occasional extended meaning, for instance *Winkel* is a laymen term for a geometrical instrument for *drawing* specific angles (E. *protractor*), but for the majority of translation purposes the translational equivalences listed above can be regarded as one-to-one in both language directions. A study of dictionary arrangements is not necessary here and such terms appear in Volume 2 only for completeness.

4.2 Dual Equivalence

As soon as terms like *Geschwindigkeit* (speed, velocity) or *Spannung* (stress, tension) appear in a text the translator's problems begin. Two approaches are adopted in the second volume to help readers faced with such dilemmas.

First, the Polyseme Dictionary lists samples of adjacent compounds at entries like *Geschwindigkeit*, *Spannung* so that the translator realises the terms are polysemous:

Fluchtgeschwindigkeit	escape velocity
Reisegeschwindigkeit	cruising speed
Schallgeschwindigkeit	speed of sound
Oberflächenspannung	surface tension
Wärmespannung	thermal stress
Zugspannung	tensile stress

Secondly, the Thesaurus provides brief "definitions" of English terms which help the reader nail down the appropriate concept. For instance:

speed	t: scalar quantity; tu: ms^{-1}, km/h, mph;
velocity	t: vector quantity; d: speed in a given direction;
stress	d: force acting per unit area; tu: newton.m^{-2};
tension	t: force; tu: newton;

where the thesaurus descriptors *t, d, tu* require substitution of the sequences *a type of, designates, typical unit* respectively. Hence the semantic feature distinguishing *speed* and *velocity* is that one term denotes a *scalar quantity*, the other a *vector*, whereas *stress/tension* are different parameters distinguished among other things by their units.

4.3 Multiple Equivalence

Some translators expect a compound term in one language (e.g. *Drehgeschwindigkeit*) to correspond to a compound in the target language (E. *rotational speed*), and are surprised that expressions like *Schubkraft* (E. *thrust*) do not. Moreover, when the concept corresponding to the compound occurs repeatedly in the source text it is inevitably shortened, and the translator has to decide whether in a particular context to translate *Kraft*, for example, as *thrust* or to insert the broader term *force*.

Such problems can occur with harmless looking expressions, such as *Abstand* (E. *distance*) which is translated by *clearance, spacing, pitch, gap* in specific contexts:

Bodenabstand	MECH	clearance
gegenseitiger Abstand	GEOM	spacing
Gewindeabstand	MECH	thread pitch
Kontaktabstand	IGN	contact breaker gap

The TPD helps specify the concepts implied by indicating possible subject fields (MECH: Mechanical Engineering, GEOM: Geometry, IGN: Ignition Systems), but cross-checking in the Thesaurus is advisable. This reveals that *distance* has a basic meaning in Physics:

distance PHYS *Abstand*
t: physical quantity; ct: time, mass, temperature, etc.

namely: a type of *physical quantity*, contrasted with concepts like *time, mass, temperature* (Chapter 1). A second entry for *distance* indicates that the reverse translation into German is not necessarily *Abstand*:

distance	MATH	*Abstand, Strecke*
braking distance	AUTO	*Bremsweg*
focal distance	OPT	*Brennweite*

The entry consists of a so-called *polyseme group* more typical of the TPD. The Thesaurus employs this lexicological device for terms like *line* (Ge. *Linie, Gerade*), *load* (Ge. *Ladung, Last, Belastung*), concepts which are easily understood but have different shades of meaning and context-specific interpretations in German:

axle load	AUTO	*Achslast*
cargo load	AERO	*Frachtladung*
dotted line	GEOM	*Punktlinie*

The polyseme group entered at *distance* indicates that, according to the context, the German equivalent could be *Strecke*, *Weg*, *Weite*, terms which are themselves polysemous:

Bahnstrecke	RAIL	rail section
Lichtweg	OPT	light path
Spannweite	AERO	wing span

This is evident in the TPD.

To obtain familiarity with the terminology of both source and target languages the reader is advised to conduct regular lexicological exercises of this nature, using the TPD and Thesaurus. The latter are not dictionaries in the ordinary sense of the word at all but organised didactic guides specifically designed for resolving semantic problems of this type in technical translation.

4.4 Polyseme Groups

When the number of entries in the TPD at a particular polyseme becomes excessive, the compounds are arranged into separate groups. For instance, the entries for *Spannung* are arranged into a group designating electrical concepts (i.e. *voltage*, *potential*, *emf*, etc.) and one denoting mechanical ones (e.g. *stress*, *tension*). The slightly longer accessing time required to locate a particular compound (if two groups of entries are scanned alphabetically instead of just one) is compensated by the didactic organisation.

Hence, the term *Kraft*, discussed in Chapter 1 and mentioned above is divided into two distinct polyseme groups — those designating a type of *force* and denoting a measurable parameter or theoretical concept, such as:

Anziehungskraft	*ELSC*	attractive force
Auftriebskraft	*AERO*	lift
Auftriebskraft	*NAUT*	buoyancy
etc.		

and those denoting a different physical quantity or something else entirely:

Bremskraft	BRAK	braking power
Kernkraft	NUCL	nuclear power
Spannkraft	MAT	elasticity
Wärmekraft	POW	heat

4.5 Hierarchic Dimensions

The German expression *Kraft* is very difficult to specify in any form of dictionary, even though the concept *force* is fundamental to almost every branch of engineering. The reason is that hyponyms exist in English which are absent in German.

After studying Figure 1 (Chapter 1) readers familiar with semantic hierarchies may deduce the following hyponym arrangement for the concept *force*:

1	force	*Kraft*
11	action	*Kraft*
12	reaction	*Gegenkraft*
13	tension	*Spannung*
14	compression	*Druckkraft*
15	friction	*Reibungskraft*
16	gravity	*Gewichtskraft*
17	thrust	*Schubkraft*
18	upward thrust	*Auftriebskraft*
181	buoyancy	*Auftriebskraft*
182	lift	*Auftriebskraft*
19	traction	*Zugkraft*

At first glance this seems a neat convenient organisation, but a closer look at the hyponyms in the TPD indicates that the hierarchy above is an oversimplification. Terms like *action/reaction* belong to a more fundamental area of science than *tension/compression* yet neither this nor the contrasts themselves are marked; *friction/gravity* are forces occuring in nature over which the technologist has little control, in contrast to *thrust/traction* which involve controlled motion. *Upward thrust* covers concepts such as *buoyancy* (Nautical Engineering) and *lift* (Aeronatics) but is not a hyponym of *thrust* itself.

Rather than leave the reader totally perplexed at this stage, Figure L1A divides the main interpretations of *Kraft* into subcategories, an arrangement equivalent to a *multi-dimensional* hierarchic structure. An appropriate field code (e.g. MECH: Mechanical Engineering) is given for each concept listed, and brief contextual explanations of the subcategories (e.g. "Electrostatics: interactions between charged particles") help the reader to specify the concepts more exactly.

The essential features of the glossary are reproduced below.

Figure L1A. Different Types of Force

A demonstration model of a multidimensional hierarchic system with terminology centred on the engineering concept *force* and the German expression *Kraft*. It employs the following dictionary labels:

ACU	Acoustics	CONS	Construction Engineering
AERO	Aeronautics	MECH	Mechanical Engineering
ASTR	Astronomy	NAUT	Nautical Engineering
ELSC	Electrostatics	NUCL	Nuclear Engineering
FLUD	Fluid Dynamics	PHYS	Physics
a:	associated with	u:	used in connection with
d:	designates	tu:	typical unit

Subcategories of Kraft

1. *Engineering Science*, d: physical quantity distinct from *power*, *energy*, etc.

Kraft (tu: newton)	PHYS	force

2. *Newtonian Mechanics*, u: fundamental properties of all forces.

Kraft	PHYS	action
Gegenkraft	PHYS	reaction

3. *Electrostatics*, u: interactions between charged particles.

Abstoßungskraft	ELSC	repulsive force
Anziehungskraft	ELSC	attractive force

4. *Mechanical Engineering*, u: forces producing motion.

Antriebskraft	MECH	propulsive force
Hubkraft	MECH	lifting force
Zugkraft	MECH	traction

5. *Mechanical Engineering*, u: quantities with different *units* (i.e. not *forces*).

Drehkraft (tu: newton.metre)	MECH	torque
Hebelkraft (tu: newton.metre)	PHYS	leverage

6. *Aeronautics, Fluid Dynamics, Shipbuilding*, u: motion or flight.

Auftriebskraft	AERO	lift
Auftriebskraft	FLUD	upward thrust
Auftriebskraft	NAUT	buoyancy
Gewichtskraft	PHYS	weight
Luftwiderstand	AERO	drag
Reibungswiderstand	FLUD	drag
Strömungswiderstand	NAUT	drag
Schubkraft	AERO	thrust

7. *Civil Engineering*, u: forces resulting in material stresses.		
Druckkraft (u: bar, beam)	CONS	compression
Spannung (u: general context)	CONS	tension
Spannkraft (u: cable)	CONS	tension
Zugkraft (u: bar, beam)	CONS	tension
8. *Physics*, u: concepts relating to Mechanics, Acoustics.		
Spannkraft	PHYS	tension
Federkraft	PHYS	spring tension
Saitenspannkraft	ACU	string tension
9. *Physics*, a: forces difficult to control.		
Reibungskraft	PHYS	friction
Gewichtskraft	PHYS	gravity
Gravitationskraft	ASTR	gravitation
Schwerkraft	ASTR	gravitation
10. *Collective Term*, a: forces, difficult to distinguish individually.		
Verbiegungskräfte	MECH	bending forces
Verdrehungskräfte	MECH	twisting forces, torsion
Kernbindungskräfte	NUCL	nuclear (binding) forces

Nevertheless, even this arrangement does not cover all interpretations of the polyseme, and it is not a practical organisation in a large dictionary anyway in view of the increased access time. The TPD therefore adopts a compromise solution and lists these entries in alphabetic order as a collective polyseme group. It was pointed out in the Introduction that often the worst errors made by technical translators involve the most fundamental engineering concepts. *Kraft* is one example. Others will follow. Simultaneous systematic study of the engineering chapters, the Thesaurus and TPD will help the reader to eventually overcome these problems.

Chapter 2 and Figures 2A–D introduce another basic technical polyseme, the German expression *Widerstand*, whose meanings denote different electrical concepts: *resistance, resistor, resistivity, reactance, impedance, rheostat*, etc. Here, the TPD employs a multidimensional internal entry arrangement similar to the arrangment for *Kraft* in Figure L1A. These arrangements are discussed later, when more engineering concepts and relevant terminology have been covered. An alternative approach to polyseme differentiation may clarify the situation at this stage: a comparison between *hierarchic listing* and the *thesaurus approach.*

4.6 Hierarchic List, Thesaurus

The hierarchic term lists of Chapter 2 are concise lexicological arrangements for relating technical concepts to broader concepts. This organisational principle can be used in specific engineering areas to simultaneously portray a large number of conceptual interrelationships, distinguished according to their *logical* (i.e. hierarchic) and *ontological* (i.e. semantic) properties by the descriptors *generic, partitive, metric, associative.* The Technical Thesaurus employs other descriptors to relate engineering concepts.

The lack of immediate hierarchic information in a thesaurus is compensated by the easier access to terminology (alphabetic order) and the fact that greater detail is possible in thesaurus definitions. Figure L1B demonstrates applications of the thesaurus descriptors to the main terminology of Figures 2A–D (Chapter 2). Readers are invited to examine the contents and compare the thesaurus definitions to the information deductible from the term lists concerned, and from that presented in Unit 3 of this handbook. Once the demonstration model is understood, use of the main Thesaurus and of larger more complex hierarchic arrangements introduced in later chapters follows naturally.

Figure L1B. Thesaurus Representation of Figure 2A–D

a:	associated with	m:	a (measurable) parameter of
co:	consist(s) of	p:	part of
ct:	contrasted with	s:	synonym for/abbreviation of
cv:	covers the concept(s)	t:	a type of
d:	defined as/designates	tu:	typical unit
ex:	(typical) example	u:	used in connection with

ac voltage (s: alternating voltage)	*Wechselspannung*
ac current (s: alternating current)	*Wechselstrom*
amplitude (u: oscillation, wave, waveform)	*Größe*
amplitude (cs: peak amplitude)	*Scheitelwert*
RMS amplitude (cs: RMS value)	*Effektivwert*
capacitance (t: electrical parameter)	*Kapazität*
capacitor (t: circuit component)	*Kondensator*
ceramic capacitor	*Keramikkondensator*
electrolytic capacitor	*Elko*
parallel-plate capacitor	*Plattenkondensator*
charge (tu: coulomb)	*Ladung*
choke (t: electrical component)	*Drosselspule*
coil (u: transformer)	*Wicklung*
conductance (tu: mho)	*Leitfähigkeit*
conductivity (tu: mho.cm^{-1})	*spezifische Leitfähigkeit*
current (tu: amp)	*Strom*
alternating current	*Wechselstrom*
direct current	*Gleichstrom*

dc current (s: direct current)	*Gleichstrom*
dc voltage (s: direct voltage)	*Gleichspannung*
dielectric (p: capacitor)	*Dielektrikum*
emf (t: voltage; u: transformer)	*Spannung*
frequency (u: waveform; tu: Hz)	*Frequenz*
angular frequency (tu: radian.s^{-1})	*Kreisfrequenz*
heat (t: energy; tu: joule)	*Wärmeleistung*
impedance (t: parameter; tu: ohm)	*Scheinwiderstand*
inductance (t: parameter; tu: henry)	*Induktivität*
inductor (t: circuit component)	*Spule*
iron core (p: inductor, transformer)	*Eisenkern*
loss resistance (u: coil; tu: ohm)	*Verlustwiderstand*
number of turns (u: winding)	*Windungszahl*
peak value (u: wave amplitude)	*Scheitelwert*
phase (t: circuit parameter)	*Phase*
phase angle (u: waveform; tu: degree)	*Phasenwinkel*
phase shift (u: impedance)	*Phasenverschiebung*
phasor (t: vector)	*Zeiger*
voltage phasor	*Spannungszeiger*
current phasor	*Stromzeiger*
phasor diagram (u: circuit design)	*Zeigerdiagramm*
plate (p: capacitor)	*Kondensatorplatte*
potential (t: voltage)	*Potential*
potentiometer (t: potential divider)	*Potentiometer*
power (tu: watt)	*Leistung*
reactance (t: parameter; tu: ohm)	*Blindwiderstand*
capacitative reactance	*kapazitiver Widerstand*
inductive reactance	*induktiver Widerstand*
reactive volt-amperage (tu: ohm)	*Blindleistung*
resistance (u: impedance)	*Wirkwiderstand*
resistance (tu: ohm)	*Widerstand*
resistance (s: electrical resistance)	*ohmscher Widerstand*
resistivity (tu: ohm.cm)	*spezifischer Widerstand*
resistor (t: circuit component)	*Widerstand*
rheostat (t: laboratory component)	*Schiebewiderstand*
RMS value (ct: peak value; u: wave)	*Effektivwert*
RMS voltage (m: alternating voltage)	*effektive Spannung*
RMS current (m: alternating current)	*effektive Stromstärke*
time constant (u: capacitor discharge)	*Zeitkonstante*
transformer (t: circuit component)	*Transformator*
turn (u: winding)	*Windung*
turns ratio (m: transformer)	*Windungsverhältnis*
volt-amperage (u: impedance; tu: VA)	*Scheinleistung*

voltage (tu: volt)	*Spannung*
direct voltage	*Gleichspannung*
alternating voltage	*Wechselspannung*
wattage (u: heating device; tu: watt)	*Wattleistung*
wattage (u: impedance; tu: watt)	*Wirkleistung*
winding (p: transformer; t: coil)	*Wicklung, Spule*
primary winding	*Primärwicklung*
secondary winding	*Sekundärwicklung*

Unit 5

Materials Science

The third disk chapter completes the reader's basic technological education, introducing a subject whose terminology, like that of its predecessors, reappears in virtually all branches of engineering but originates from *Chemistry* as well as from *Physics*: the field of *Materials Science*. The chapter begins with a brief examination of the engineering implications of the term *material* and then discusses one of the main tools of the materials scientist: *the Periodic Table of Elements*. A knowledge of the Periodic Table enables technologists to understand how atoms combine to form *molecules*, *compounds* and ultimately *materials*, and how a vast range of material *properties* occur — mechanical, optical, electrical, magnetic, thermal, chemical, radioactive. It also enables them to design *new* materials with specific properties *engineered* to specific applications. The chapter has the following main sections:

Life for the reader gets easier in this area. The chapter begins by introducing and differentiating a small set of *chemical terms*, such as *mass number*, *group number*, *valency*, *isotope*, *nuclide*, that reappear frequently elsewhere in the book, especially in Nuclear Engineering (Chap. 4) and in the *polymer* and other *chemical industries* (Chap. 10). It then reveals a good example of a *technical verb* which is polysemous in German: *binden* (E. *bind*, *bond*). It is conjugated either *bind/bound* or *bond/bonded*, according to the context. The reader will encounter other technical verbs later in the book, and indeed these are one of the features of the third major didactic glossary, the *Technical Collocation Dictionary* (TCD).

5.1 Material Properties

Whether they are designing bridges, computers, power stations or space vehicles, all engineers require an intimate knowledge of the *materials* available to them and their respective *properties*. Even the manufacture of a simple automobile requires a diverse range of materials, *iron*, *steel*, *glass*, *rubber*, *plastics*, not to mention the variety of electrical and electronic components, and for *steel* alone there are at least 2000 varieties to choose from. Different *material strengths*, *densities*, *electrical* or *thermal conductivities* have to be taken into account, as well as considerations such as:

i. *machinability* — whether the material is *machinable*, i.e. can be cut or shaped using *machine tools* (Ge. *Werkzeugmaschine*);
ii. *formability* — whether it can be shaped or *cast* in a hot molten state in a *mould* (Ge. *Form*);
iii. *durability* — whether the material loses its strength or is subject to corrosion or other chemical decomposition as it ages;

and many other factors. This chapter draws attention to some of the *parameters* employed by engineers when selecting materials for specific design applications.

Different technologists are interested in different *material properties*. A mechanical engineer may be interested in the *stresses* and *strains* to which a particular type of *perspex* (Ge. *Plexiglas*) may be subjected whereas the designer of a greenhouse is more interested in its *optical* and *thermal* parameters, how much *light* it lets through and how much *heat* it retains. Similarly, an electrical engineer requires uniform materials with a constant specific *conductivity*. The conductivity may be high in the case of *conducting materials* but should be negligibly low for *insulators* or *dielectrics.* Electronic engineers require special *semiconductive materials* (Ge. *Halbleiter*) with combinations of these properties. Just as automobile engineers are perpetually producing innovative designs for new vehicles manufactured from the same basic set of machinery, so scientists are perpetually designing new materials by rearranging or establishing new combinations of the basic atoms.

5.2 Mechanical Properties

Gears, drive components and other parts of the transmission system of a motor vehicle require materials which can be easily machined in the initial manufacturing stages and then strengthened in the final production stages to withstand rigorous everyday usage. On the other hand, *bumpers* (Ge. *Stoßstange,* Am. *fender*) have to be made from materials which are easily shaped but resist deformation on impact. The chapter looks closely at what is meant by *material strength* and how it relates to other factors such as *elasticity, ductility, creep, hardness.*

Chapter 1 draws a distinction between the concepts *stress* and *strain* (Ge. *Spannung, Dehnung*). For example, a metal bar of length 1 metre, subjected to a heavy load producing a *stress* of 500 kilonewtons/square-metre (500 kNm^{-2}), stretches by 0.25 mm, thus resulting in a proportional elongation of 0.00025, the so-called *strain.* When a material is *stressed,* in other words stretched, compressed or deformed in some other way, but regains its shape and returns to its original state as soon as the stress is removed, it is said to have been subjected to *elastic strain.* Elastic strain is directly proportional to the stress applied, and the ratio of the two quantities (*stress* divided by *strain*) is known as the *modulus of elasticity.* When the *tension* or *compression* is increased beyond the *elastic point* (Ge. *Elastizitätsgrenze*), the material undergoes *plastic deformation.* Plastic strain entails permanent deformation. *Strain* is no longer proportional to *stress,* and the design engineer requires a special table or graph of the relationships between the two quantities, the *stress/strain chart.*

The chapter introduces a broad range of terminology, expressions like *yield stress, breaking strength, hardness, toughness, creep,* analysing the *parameters, units* and their German equivalents. These terms, especially the parameter *creep,* which concerns changes in stress/strain behaviour when a material *ages* or is subjected to persistent *loads* for long periods, are important in Construction Engineering and reappear in Chapter 13.

5.3 Chemical, Electrical, Thermal Properties

The electrical property of main interest to design engineers is *conductivity* (Ge. *Leitfähigkeit*) and its behaviour at different *temperatures*. Conductivity is measured using the SI unit $ohm^{-1}.cm^{-1}$ and is closely related to the *resistivity* of the material (Ge. *spezifischer Widerstand*; unit: *ohm.cm*). Other basic electrical parameters, *resistance, reactance, impedance*, etc., come into play at a more advanced stage in the design of electrical equipment than that of *ordering* the materials, but a large section of Materials Science is devoted entirely to one particular area of electrical engineering: the design of *semiconductor materials* for fields like Electronics or Data Processing.

There is a parameter analogous to *electrical conductivity* used in the description of the *thermal properties* of materials: *thermal conductivity* (Ge. *Wärmeleitfähigkeit*). Just as *electrical conductivity* is a measure of the rate of *charge flow* per unit of electrical energy (voltage) applied, so *thermal conductivity* expresses the rate of *heat flow* to the *heat applied*. Like every other form of energy, *heat* is now measured almost universally in the unit *joule*. The *calorie* received an official funeral in the seventies and units like the *Btu* (British Thermal Unit) are employed only in very specific areas.

Regarding the *chemical properties* of materials, the interest of most designers centres on their resistance to *corrosion*, especially to *oxidation* in the presence of water, the phenomenon normally known as *rusting*. Various parameters are employed to describe this, for instance *millimetres of surface lost per year*. This area is related to Metallurgy (Chapter 9). A second separate area of engineering has evolved almost entirely from the study of Materials Science, the *polymer industry* (Chapter 10).

5.4 Lexical Gaps

An interesting aspect of Chapter 3 from the linguistic viewpoint is the demonstration of the fact that technical terms present in one language can be completely absent in another, when the *concept* itself is absent. There seem to be no exactly equivalent expressions in technical English for two German materials science terms: *Atomrumpf, Atomhülle*. The concepts can only be described.

Any atom which has become separated from one of its *valence electrons* becomes a *positive ion*. Various terms denote this same entity in different engineering contexts: *cation, ion core, ionised atom, host atom*. Each of these terms could correspond to the German concept *Atomrumpf.* For instance, the ions present in *solids* generally remain at fixed *sites* in the material; only the electrons move. Thus, the terms *ion, positive ion, positive atom, parent atom, host atom, fixed atom,* which occur throughout the literature, denote different attributes of what is essentially the same German concept: *Atomrumpf.* Yet English has no exact designation for this concept.

Selection of the "correct" English term in a context like the above is a matter of common sense. Substitution of common but very misleading dictionary suggestions, such as *"atomic residue", *"atomic trunk", must be avoided at all costs, as these do not appear in the normal technical literature and this makes translations difficult or impossible to understand. This is rather like the situation with another dictionary coinage *"inductivity", mentioned earlier (Unit 3). The fact that it might appear in a dozen different bilingual dictionaries does not necessarily mean that the term is correct, simply that an entire generation of lexicographers have copied one another's mistakes.

Like *Atomrumpf,* the German *Atomhülle* (frequent synonym: *Elektronenhülle*) constitutes a similar problem for translators. It implies the set of *electron shells* (Ge. *Elektronenschale*) which surround the nucleus of an atom in the manner in which the atmosphere, stratosphere and ionosphere surround the earth. Once again, dictionary suggestions like *"atomic sleeve", *"atomic envelope", *"atomic mantle" have no basis in normal engineering practice. English-speaking technologists do not employ the concept, but translation problems can sometimes be avoided simply by rendering *Schale* as *shell* and *Hülle* as *shells.*

Lexical gaps occur in general language, and the more remote the language relationship (English and Native American languages) the more likely their occurrence. Some linguists may be surprised to find that they also occur in the *technical languages* of very similar industrialised countries.

5.5 Microthesaurus Construction

Like its predecessors, the third chapter presents a selection of microglossaries providing German equivalents to terminology employed and illustrating a variety of conceptual inter-relationships. Figure 3D provides a didactically ordered *microthesaurus* too.

The *descriptors* (*d:* "designates/denotes", *u:* "used in connection with", *p:* "part of", etc.) are the same as those of the main Technical Thesaurus of

Volume 2, but the structural principles are slightly different. The microthesaurus is reproduced below, so that the reader can compare the two approaches at his or her leisure.

Figure 3D. Microthesaurus of Materials Science Expressions

atomic number (d: number of protons in an atom)	*Kernladungszahl*
atomic mass (d: mass of an atom; tu: amu)	*Atommasse*
coefficient of expansion (t: thermal parameter)	*Ausdehnungskoeffizient*
coefficient of linear expansion	*Längenausdehnungskoeffizient*
coefficient of volume expansion	*Volumenausdehnungskoeffizient*
conduction (a: heat or current flow)	*Leitung*
conductivity (t: parameter; u: conduction)	*Leitfähigkeit*
electrical conductivity (a: resistivity)	*elektrische Leitfähigkeit*
thermal conductivity	*Wärmeleitfähigkeit*
deformation (d: change in the shape of an object)	*Verformung*
elastic deformation	*elastische Verformung*
permanent deformation	*bleibende Verformung*
plastic deformation	*plastische Verformung*
disintegrate (u: nuclear fission)	*zerfallen*
doping (u: impurity injection; a: semiconductors)	*dotieren*
elastic point (d: limit of elastic deformation)	*Elastizitätsgrenze*
electrolysis (u: current conduction in electrolytes)	*Elektrolyse*
electrolyte (t: solution containing mobile ions)	*Elektrolyt*
electron shell (p: atom)	*Elektronenschale*
(set of) electron shells	*Elektronenhülle*
electron cloud (u: metals, metallic bonding)	*Elektronengas*
element (t: fundamental substance; ct: compound)	*chemisches Element*
inert element (t: highly stable, non-reactive element)	*Edelelement*
noble element (ps: inert element)	*Edelelement*
metallic element (cs: metal; ex: calcium, iron, mercury)	*Metall*
non-metallic element (ex: sulphur, arsenic, chlorine)	*Nicht-Metall*
group (a: Periodic Table; ex: Group IV)	*Gruppe*
isotope (d: element with more neutrons than the basic type)	*Isotop*
mass number (d: number of nucleons in an atom)	*Massenzahl*
modulus of elasticity (d: ratio of stress to strain)	*Elastizitätsmodul*
Young's Modulus (u: linear stress/strain curve)	*Elastizitätsmodul*
nucleon (p: nucleus; cv: proton, neutron)	*Nukleon*
nuclear forces (a: nucleons inside a nucleus)	*Kernbindungskräfte*
nuclide (d: one of several isotopes of an element)	*Nuklid*
positive ion core (d: atom lacking an electron)	*Atomrumpf*
resistivity (t: electrical parameter; tu: ohm.mm)	*spezifischer Widerstand*
semiconductor material (a: easily varied conductivity)	*Halbleiter*

strain (d: deformation due to stress)	*Dehnung*
strain (t: parameter; tu: dimensionless)	*Dehnung*
stress (u: forces leading to deformation)	*Spannung*
stress (t: parameter; tu: newton.m^{-2})	*Spannung*
subgroup (p: group; a: incomplete inner shell)	*Nebengruppe*
valency (a: element; d: number of outer shell electrons)	*Valenz*

Unit 6

Nucleonics

Chapter Four adopts a different approach to that of its predecessors. Instead of describing key concepts in detail and later examining the terminology for possible translation difficulties, the initial description centres on terminology itself. From this point onwards the book moves away from Science and into the realm of Engineering. Increased specialisation means that descriptions become more concise, and the *micro-glossaries* and *microthesauri* closing the chapters become more detailed and much larger. The chapter concerned has the following structure:

- 4.1 Radiation, Radioactivity
  - 4.1.1 Radio-Chemistry, Radio-Astronomy
  - 4.1.2 Radiant Energy, Harmful Radiation
- 4.2 Alpha, Beta, Gamma Radiation
  - 4.2.1 Particulate/Electromagnetic Radiation
  - 4.2.2 Frequency, Wavelength, Distance
  - 4.2.3 Acoustic, Electromagnetic Waves
  - 4.2.4 Energy Quantum, Photon
  - 4.2.5 Wave/Particle Duality, Diffraction
  - 4.2.6 Rem, Becquerel, eV, MeV
  - 4.2.7 Beam, Ray, Current, Stream
- 4.3 Radio-Element, Radio-Nuclide, Radio-Substance
  - 4.3.1 Radiant, Radioactive, Radiological
  - 4.3.2 Decay, Disintegration, Decomposition, Dissociation
- 4.4 Matter, Anti-Matter
  - 4.4.1 Positron, Neutrino, Quark
- 4.5 Fission, Fusion, Decay
  - 4.5.1 Nuclear Power Generation
- 4.6 Nuclear Waste Disposal
  - 4.6.1 Reprocessing Plant, Repository
  - 4.6.2 Landfill, Disposal Site, Dump

The chapter begins with a close look at distinctions between *radiation* and *radioactivity* from the common viewpoint of a nuclear scientist, radiochemist or radiological waste-disposal authority. It discusses the various semantic interpretations of the morpheme *radio* in expressions like *thermal radiation*, *visible radiation*, *gamma radiation*, and the linguistic relationships to common everyday expressions like *radio set*, *radiator*, *radiotherapy*, despite the apparent absence of any direct semantic relationship (cf. Ge. *Rundfunkempfänger*, *Heizkörper*, *Strahlentherapie*). A clear insight into the *etymology* of the technical language of Nucleonics assists the reader to understand more complex specialised concepts such as *radionuclide*, *positron radiation*, *radiological accident* in later subsections.

Like its predecessors, the chapter discusses:

i. *basic conceptions* necessary for a broad understanding of the subject in both English and German — *fusion*, *diffraction*, *spent fuel*, *terminal waste*, *radiation dosage*;
ii. important *parameters* appearing repeatedly throughout the field, *wavelength*, *transition period*, *decay rate*, *half-life*;
iii. any units peculiar to the field, *rem*, *bequerel*, *Mev*;
iv. *contrasting expressions* — *acoustic/electromagnetic wave*, *particulate/electromagnetic radiation*, *negatron/positron*;
v. areas of semantic overlap that can lead to *contextual synonymy* — *photon/energy quantum*, *disposal site/repository*;
vi. *non-countable* technical expressions (NCNs), *matter*, *anti-matter*.

Indeed these aspects of technical language are featured in *all* the remaining chapters.

6.1 Nucleonics, Nuclear Engineering

Nucleonics concerns the detailed study of *atoms*, *nuclei* and their constituents, so-called *nucleons*, as well as the *radiation* and other *energy* released when nuclei are disrupted or fused together. Its development began with the increased use of *nuclear technology* for both peaceful and cold-war military purposes in the early nineteen-sixties, and it is still evolving in different directions as a twenty-first century *applied science*. The main industrial applications lie in the field of *nuclear power generation*, and there are other practical applications in areas as diverse as *Archaeology* (radiocarbon dating), *Astronomy* (studies of distant galaxies and supernova), and *Medical Science*, especially the area of Radiological Diagnostics (Ge. *nuklearmedizinische Diagnostik*), where supplies of *radioisotopes* to hospitals are almost as regular as supplies of synthetic rubber to an automobile plant.

Nucleonics is an area in which some translators initially feel out of their depth. The chapter introduces simply the main features and basic terminology underlying the discipline. Though there is no direct link with the preceding chapter on Materials Science, a connection with Basic Chemistry means that certain previously discussed conceptions reoccur with similar connotations, expressions like *atomic number*, *mass number*, *isotope*, *nuclide*, *elementary particle*, *nucleon*. Chemistry deals with *interactions* between substances, but there is another field more closely related to Nuclear Engineering, which deals with the *decay* or *disintegration* (Ge. *radioaktiver Zerfall*) of substances: Radio-Chemistry. The two areas examine similar features of nuclear reactions but from differing aspects: Nucleonics from the viewpoint of the engineer or physicist, Radiochemistry from the viewpoint of the general scientist or chemist.

6.2 Radiation, Radioactivity

Rather like their German counterparts *Strahlung/Radioaktivität*, the terms *radiation* and *radioactivity* have ominous connotations in general colloquial English and, at first sight, they appear to be contextually synonymous. To level-headed scientists, the terms have quite different meanings and are without the disturbing overtones.

The basic meaning of *radiation* (Ge. *Strahlung*) concerns the mechanism whereby *energy* is transferred over large distances by means of "energy rays" or more specifically: *electromagnetic waves*. In the Physics branch known as *Heat Transfer* (Ge. *Wärmeübertragung*), the expression *thermal radiation* (Ge. *Wärmestrahlung*) contrasts on the one hand with *conduction* (Ge. *Wärme-leitung*) and on the other with *convection* (Ge. *Wärmeströmung, Wärme-konvektion*). *Heat* is a form of energy, and so is *light*. This technical meaning of *radiation* is extended to imply the *electromagnetic waves* themselves which carry the *radiated energy*. Thus *light* is classified as *visible radiation. Radio waves, microwaves, ultra-violet rays, X-rays* and *gamma-rays* constitute different types of *invisible radiation. Heat* arriving from the sun can be considered as an effect resulting mainly from *infrared radiation*.

Some substances *emit* radiation and in so doing *disintegrate*, forming other substances. The activity or process of emitting radiated energy is termed *radioactivity*. Just as the narrow meaning of *radiation* (a means of *energy transfer*) also covers *waves* carrying *radiated energy*, so the technical meaning of *radioactivity* is extended to cover the *energy* itself and contrasts with other more familiar energy forms such as *heat, light, sound, mechanical energy*. Radiation which causes substances to *ionise* is known in German engineering circles as *radioaktive Strahlung*. English-speaking nuclear technologists employ the term *ionising radiation*, and normally reject both the literal mis-translation *"radioactive radiation" and the expression coined from general language: *"radioactivity".

6.3 Radio Morphology

At this stage, the reader is invited to take a closer look at various semantic interpretations of the morpheme *"radio(–)"*, as a prefix in expressions like *radio-element, radiochemist, radioastronomer* and in the terms *radiation, radioactivity*. The two distinct basic meanings of the prefix *radio* are easily clarified:

i. In expressions like Radiochemistry or Radiology, the morpheme implies a direct connection with *radioactivity* or *ionising radiation*. This is not true of Radioastronomy. The latter concerns the study of distant *celestial bodies* (Ge. *Himmelskörper*) whose energy reaches us in the form of *radio waves* as opposed to visible light.
ii. A second prefix *radio-* appears in terms like *radiosubstance, radionuclide, radioisotope*. This is just a technical abbreviation of the concept *radioactive*. Thus the term *radio-element* merely implies *radioactive element*.

Throughout Nucleonics, and its associate fields Radiochemistry and Nuclear Power Technology, the narrow meanings of *radiation* and *radioactivity* specified above are in fact *broadened*. They imply not just energy radiated by electromagnetic waves but also that carried by specific fast-moving *particles*, among which are *alpha* and *beta particles*. This second type of *radiant energy*, known as *particulate radiation*, is discussed in Section 4.2 (of the disk) along with its counterpart *electromagnetic radiation*.

6.4 Radiant Energy, Harmful Radiation

Normally when technologists in one country use the same expression in different ways or in different fields it causes headaches for non-technically minded translators as there are different L2 equivalents for each polyseme. Terms like *Impuls* (impulse, pulse, momentum), *Widerstand* (resistance, reluctance, reactance, impedance), *Spannung* (tension, stress, voltage, potential), all of which occur in Basic Mechanical or Electrical Science, demonstrate this phenomenon conclusively. Yet *Strahlung*, despite its very different implications and connotations in expressions like *Wärmestrahlung*, *Kernstrahlung*, *elektromagnetische Strahlung* seems to be always translated by *radiation* and vice versa. This is not *one-to-one equivalence* (Unit 4) but *parallel polysemy*.

There is no translation problem this time, but to avoid misunderstandings in the chapter and in general in this field, readers may find the following summary useful, a brief illustrative list of the main alternative technical interpretations of the term *radiation*:

i. a means by which thermal energy is transferred (ct: *convection*, *conduction*);
ii. electromagnetic waves constituting transferred energy (ex: *visible radiation*, *microwave radiation*);
iii. dangerous electromagnetic rays causing the ionisation of tissue and other substances (ex: *gamma radiation*, *X-rays*);
iv. dangerous particles causing ionisation (ex: *alpha radiation*);
v. collective term for mixtures of electromagnetic and particulate radiation resulting from a nuclear reaction (cs: *nuclear radiation*).

6.5 Beam, Ray, Current, Stream

Many disk users will not be English native-speakers, but *German* translators into or from English. For their benefit, and also to provide other readers with a brief respite from the bombardment with scientific information, the disk examines difficulties arising in connection with polysemous terminology, which would not necessarily trouble *native* English speakers. Interest here is concentrated on the German polysemes *Strahl* (E. *beam*, *jet*, *ray*, *stream*), *Strom* (E. *current*, *stream*).

For instance, in a context involving the *picture tube* (Ge. *Bildschirmröhre*) of a TV set or monitor screen, German translators may wonder why *Lichtstrahl* is rendered as *light ray* whereas *Elektronenstrahl* is translated as *electron beam*. In a different context, one involving what British speakers call a *torch* (Am. *flashlight*, Ge. *Taschenlampe*) the term *Lichtstrahl* is more likely to be translated by *beam*. This problem is easily resolved.

Even in general language a *ray of light* has connotations of a single light wave possibly passing through an aperture or gap (leaves, curtains, etc.), whereas a *light beam* is something much brighter. In technical language, the term *beam* is used when the individual *rays* are concentrated at a *focal point* and obliged to move in a parallel direction. Hence the expressions: *electron beam*, *torch beam*, etc. When the context concerns not waves but particles or near-particles moving in the same general direction, the correct term may be *stream*, for instance a *stream of photons, neutrons, alpha-particles*. The expression *jet* is not appropriate in the contexts discussed. It is employed in connection with liquids: *water jet, spray jet, ink jet* (Ge. *Wasser-, Sprüh-, Tintenstrahl*).

The second German polyseme *Strom* is rarely problematic. *Plasmastrom* may imply:

i. *plasma current* — an electric current resulting from *ionised particles*;
ii. *plasma stream* — ionised particles travelling along a particular path or *trajectory* (Ge. *Bahn*), often a circular one;
iii. *plasma flow* — when attention is focussed upon the *flow rate, density* or *velocity* of the particles.

Generally speaking, *Strom* should only be translated as *current* when the implication corresponds to the normal electrical significance (Ge. *elektrischer Strom*). In other cases, different equivalents should be sought or the translator must resort to paraphrasing, if confusion is to be avoided.

It is interesting to note that there is an intuitive conceptual distinction between *ray* and *wave*, as in *light ray* and *light wave*, but no physical one. This is apparent in the *electromagnetic spectrum* too (Figure 4A), which employs expressions like *radio wave, light wave, X-ray, gamma-ray* for what is essentially the same entity: *radiation*. Fortunately, German terminology follows the same knowledge patterns here as English; it uses terms like *Gamma-Strahlen* rather than *"Gamma-Wellen".

6.6 Decomposition, Disintegration, Dissociation

It is evident from the above that the disk provides not just a powerful shortcut to the acquisition of professional skills needed by translators, it also provides university academics involved in the *training* of student translators with a powerful insight into the conceptual problems encountered, especially by German students. Technical translation is an activity that requires linguists to perpetually differentiate *concepts* and narrow down *terms*. Non-native speakers are at a greater disadvantage in this respect, but even native speakers can make a total mess of a translation by employing a slap-dash approach to general technical terminology. Consider, for instance, the German expressions *Zerfall*, *Zersetzung*, which are translatable in various contexts by the terms: *decay*, *disintegration*, *decomposition*, *dissociation*.

Decomposition and *dissociation* are unrelated to Nucleonics. The terms belong to fields like Metallurgy or Chemical Engineering: a *steel structure* may *decompose* (rust and become dangerous) when exposed to extreme weather conditions; *sulphurous acid* (H_2SO_3) *dissociates* into *water vapour* and *sulphur dioxide* when heated or allowed to dry up. Only *decay*, *disintegration* are relevant to Nuclear Science, and even they are not synonyms.

But the terms are close, and like *ray/wave*, *gravity/gravitation*, *voltage/emf* and one or two other marginal terminological or semantic discrepances discussed in the early chapters, there is no *physical* distinction between *nuclear decay* and *nuclear disintegration*. Translators have to adapt their work either to the knowledge structure of the target language, or merely to what is said, as opposed to what is not said. *Disintegration* is the general expression. It is applied to nuclear transitions which take place over a wide range from several nanoseconds to a few weeks. *Decay* is the term employed when hundreds or thousands of years are involved. The period between is left to the translator's discretion.

In a text concerning the accurate *radio-carbon dating* of archaeological artefacts, the preferred term is likely to be *decay*, whereas one involving the generation of nuclear power might warrant *disintegration*. A third text, concerning the disposal of *nuclear waste*, may justify either.

6.7 Storage, Disposal

Like chemical waste disposal, the disposal of nuclear waste is an area where translators sometimes need to pay more attention to politics than semantics. Relatively straightforward, seemingly non-technical German expressions like *Endlagerung*, *Endlager*, *Zwischenlager*, which at first sight or in other contexts could be translated directly as *ultimate storage*, *final storage site*, *intermediate storage facility*, tend to have their meanings slightly disguised in English by fanciful terms like *terminal disposal*, *terminal repository*, *transitional repository* which direct public attention away from the fact that nothing is really done to extremely hazardous *spent fuel* at these sites, except to store it safely out of harm's way.

Nuclear waste itself is divided into different categories. One of these is referred to as *highly active waste* (Ge. *hochradioaktiver Abfall*), which internal reports abbreviate to *HAW*. Use of the expression *highly active* is possibly to avoid the repetitive sound of the morpheme *radio-* but more likely to avoid the ugly connotations of the full expression *highly radioactive*. Like its counterpart in the chemical industry, *nuclear waste disposal* has become a branch of engineering itself, especially *HAW disposal*. Due to the high degree of international cooperation, it requires the services of skilled translators at every level.

6.8 Reprocessing Plant, Repository

In view of the global hazard, opinions world-wide are moving towards *multinational reprocessing sites* and a carefully guarded *multinational repository* for the world's nuclear waste. Countries like Britain and Japan, which do not have convenient geological sites for waste disposal, have made considerable large-scale investments in *reprocessing facilities* and import *spent fuel* (i.e. *HAW*) from European and other countries in order to run their plants economically. The *residual nuclear waste* is usually returned to the country of origin for *terminal disposal*. The residual waste is first embedded in a special *cement* and poured into *lead containers*. Germany has a number of convenient *geological containment sites*, in *salt domes*, which absorb heat generated from radioactive waste underground without the likelihood of the occurrence of *geological fissures* and the release of hazardous radiosubstances into the environment. Other countries obtain an economic backhander from their geological advantages concerning *permanent* repository sites (Ge. *Endlagerstätte*). These import *terminal waste* from elsewhere and arrange for *permanent storage*. But sometimes waste is too dangerous for immediate shipment either before or after processing. In cases where the half-lives of the *hazardous radiosubstances* are relatively short (up to a year or so), it is safer to employ intermediate storage (Ge. *Zwischenlagerung*) at a *transitional repository* (Ge. *Zwischenlager*).

Profiting from the fact that nobody really wants *nuclear waste* on their doorstep, several enterprising *waste-management* consortiums have put forwards plans for *global repositories* to store the world's nuclear waste. Suitable geological sites have been suggested in remote desert areas of North America and Central Australia, but public pressure has been so intense that *nuclear power plants* and *military customers* are now looking elsewhere in the world for places to unload their waste. The most likely candidate at present is Russia, a country with sufficient geological facilities, the necessary intellectual expertise and above all a desperate need of foreign exchange liquidity. Contracts with Switzerland have been agreed and are in the making with other European countries too.

6.9 Landfill, Disposal Site

From the translation aspect, the concept *nuclear waste* (Ge. *radioaktive Abfälle*) can imply *spent fuel, HAW, residual, transitional* or *terminal waste* according to the context. German tends to be more direct in its terminology than English: *Entsorgung* implies *disposal* in the basic sense; *Lagerung* implies *storage* for a limited or unlimited period of time. English sometimes blurrs this semantic distinction for political rather than scientific reasons.

This phenomenon occurs elsewhere in the waste-processing industry too. The German expression *Deponie* basically means *dump*. The military expressions *ammunition dump, fuel dump* imply *storage sites* for valuable commodities essential to a military campaign. This meaning of *dump* could be extended to storage sites for different types of *waste*, but in practice it is not, owing to the connotations of a second expression *dump*, meaning a place where rubbish is *discarded* and conveniently forgotten (i.e. *dumped*). In fact the connotations of *rubbish dump* (Am. *garbage dump*) are so strong that even in a text concerning harmless non-recyclable *household waste*, the German term *Deponie* is translated not by *dump* but by *landfill*. The expression *Deponie* should no longer occur in connection with *chemical waste*, but if it does, translators might be encouraged to look closely at the source text to see whether it is possible to substitute the term *disposal site* (Ge. *Entsorgungsstelle*). *Nuclear waste* belongs either at a *reprocessing site* or at a *repository*.

Technical translators generally have little or no choice in the selection of terminology. Target-language equivalents, once chosen, are either correct or incorrect, understandable or incomprehensible, according to the degree of polysemy encountered in the source-language and the level of proficiency of the translator. The *waste disposal industries*, whether *household, chemical* or *nuclear*, are one of the few areas where technical translators still enjoy a small degree of flexibility. As competition increases, however, these industries will no doubt advance and terminologies will stabilise, in which case areas of semantic overlap in expressions like *disposal site, storage site, repository* may soon disappear, along with this current, small degree of translation freedom. Whether translators will be obliged to adapt their choices of nomenclature, to suit the customer or target reader intended, will depend to a large extent on how eagerly the public eye monitors these industrial changes.

Unit 7

Lexical Interpretation

This unit introduces the third major glossary of Volume 2, the *Technical Collocation Dictionary* (TCD), which provides access to illustrations of terminology *in context*. The unit also takes a closer look at scientific terminology encountered early in the book, and demonstrates how distinctions between different semantic aspects of this terminology are reflected in the lexical and syntactic rules concerning its usage.

7.1 Collocation Lists

The structure of the TPD itself enables the translator to rapidly distinguish between closely associated L2 alternatives:

Gehäuse	chamber, boss, housing, casing, case
Welle	shaft, spindle, rod
Zange	pincers, pinchers, pliers

as well as helping to obtain appropriate renderings for concise terms employed in specific situations:

Gerät	appliance, device, unit, equipment
Mittel	agent, coolant, flux, lubricant, solvent
Scheibe	disc, pulley, screen, wheel

But not just nouns, also specialised verbs, e.g. *disassemble, decompose, dissociate*, and specialised adjectives, e.g. *conductive, ductile, fissile, malleable*, constitute an integral part of the terminology of specific subject areas. In such cases it is useful to reveal possible contexts and demonstrate any syntactic or other adjustments necessary in translation, especially when the term has no exact equivalent in the target language and has to be paraphrased. There are *collocations* in the *TCD* that fulfil this purpose.

The main function of the TCD is to tackle problems involving general vocabulary with specific implications in engineering situations, for instance the many interpretations of the German expressions *groß*, *leicht*, *stark* in technical literature. But the dictionary copes with a number of other functions simultaneously, such as the translation of common phrases originating from *Mathematics*, the recommended use of *prepositions* in technical contexts, and the illustration of *specialised verbs* in context. *Polysemy* is evident in the TCD too, especially among nouns, for instance:

Bahn	lane, path, orbit, trajectory
Größe	amplitude, magnitude, quantity, size, value
Weg	distance, medium, means, path, route

where the same German expression appears in different engineering situations, each with its own specific target-language interpretation.

The collocation examples demonstrate typical situations in the final stages of the process of translation, when contextual and terminological problems are overcome and the translator has to provide the framework of a well-formed specialised target-language sentence. Demonstrations of the usage of *technical* terminology introduced in the engineering chapters appear too, but these are a bonus. A collocation dictionary illustrating *all* the implications, contexts and German interpretations of *all* the terminology of the engineering chapters would swamp the disk completely. Collocational arrangements are a topic for further research, but large-scale, on-line dictionaries could nevertheless soon appear if lexicographers receive unlimited first-hand access to published technical literature in machine-readable form.

The rest of this unit re-examines the distinctions between countable and non-countable technical nouns (CNs/NCNs), introduces further important noun categories, and describes how this aspect of translation is tackled in the TCD, and in the other major glossaries: the *Thesaurus* and *Technical Polyseme Dictionary* (*TPD*).

7.2 Countable, Non-Countable Nouns

Grammar books propose simple tests for distinguishing CNs from NCNs. One of these involves the possibility of co-occurence with the quantifier *some* or the article *a/an* and the existence or non-existence of a plural form. Applying this test to the terminology of Chapter 1 produces the following results ("*" indicates non-grammatical technical expressions):

i. *angle, area, density* are CNs (**some angle, *some area, *some density*);
ii. *heat, work, friction* are NCNs (**frictions, *heats, *a work*);
iii. *energy, force, power* may be CNs or NCNs according to the interpretation.

But the "*some*" test requires closer inspection and can give misleading results:

1. The vehicle has virtually stopped but still has *some velocity*.
2. *Some current* passes through the first transistor, the rest through the second.
3. *Some power* is lost.
4. *A heat of* approximately 100 J escapes through each face of the metallic enclosure.

Some velocity (ex.1) implies "a small, measurable or finite velocity", whereas *some current* (ex.2) denotes "part of the current". *Some power* (ex.3) could have the semantic implications of 1 or 2, according to the context, or it could be equivalent to the simple statement "*power is lost*". Thus there are at least three different interpretations of *some*. Example 4 proves that the NCN *heat* can appear as a CN with an indefinite article, but only in a very restricted context when the statement is part of a mathematical derivation.

There are other tests involving expressions like *amount of, number of, lot of, many*:

1. equal amounts of heat/a lot of heat
2. *"equal amounts of angle"/*"a lot of angle"
3. equal amounts of energy/a lot of energy
4. *"equal numbers of energy"
5. equal amounts of charge/a lot of charge
6. equal numbers of charges

They distinguish NCNs such as *heat* from CNs like *angle*, but results for expressions with dual properties (CN/NCN) may be inconclusive: "*equal numbers of energy*" is not possible, in contrast to "*equal numbers of charges*". The dual category (CN/NCN) thus contains subcategories of its own.

Only the plural-test provides true evidence of countability:

1. Like charges repel. Unlike charges attract.
2. Like charge repels. Unlike charge attracts.
3. Like poles repel. Unlike poles attract.
4. Like matter can coexist. Unlike matter (matter/antimatter) results in mutual annihilation with the release of energy.

The CN *charge* (ex.1) exists side by side with the NCN *charge* (ex.2). Though the connotations are slightly different, i.e. *charge* (ex.2) is likely to refer to *charged objects* (e.g. a metal bar rubbed with a soft cloth), *charge* (ex.1) denotes *charged particles* (e.g. *ions*), these are simply alternative statements of the same physical law. Example 3 provides a similar statement for the field of Magnetism as opposed to Electrostatics, but *pole* can only function as a CN. Example 4 indicates that *matter* is a NCN. Expressions like *"a matter", *"matters" (Ge. *Materie*) do not exist.

Note: Use of the specialised adjectives *like/unlike* (Ge. *gleichnamig/ungleichnamig*) is restricted to the above fields and virtually restricted to contexts similar to the above. These adjectives belong to a special grammatical category and can only be used *before* nouns. Statements like * *"the charges are like"*, *"*the poles are unlike"* are substandard.
Solution: *the charges are of the same type, the poles are opposite.*

7.3 Dual Terms, Different Terms

The dictionaries differentiate four broad classes of noun:

i. CNs: *screwdriver, electron, power plant, chassis*
ii. NCNs: *inertia, friction, gravity, heat, work*
iii. CN/NCNs, similar meanings (dual nouns): *energy*-CN (the parameter); *energy*-NCN (the concept)
iv. CN/NCNs, different meanings: *light*-CN (cs: lamp); *light*-NCN (cs: optical energy)

Nouns which are always CNs (the vast majority) are unmarked, unless it is necessary to distinguish them from similar NCNs. Entries, such as the physics parameters *friction, gravity, heat, inertia, work,* the engineering expressions *matter, equipment, information, data,* or the electronics terms *attenuation, distortion, interference* contain the descriptive label *NCN.*

The third class, dual nouns (CN/NCN), is largely restricted to concepts revolving around *physical quantities,* such as *energy, power, resistance, reactance.* The Thesaurus attempts to provide separate entries for separate concepts with different interpretations. But without overcomplicating the dictionary itself. Whether the separate entries concerned correspond to the third or fourth class of noun listed above, or even to some intermediate category, is normally evident from the *thesaurus definition.*

Nouns of the type *light* (lamp, optical energy) involve *different* terms and therefore separate entries in both the TPD and the Thesaurus. Here *polysemy*, the great enemy of technical translators, adopts a more benevolent stance: identical lexemes refering to very different concepts are easily distinguishable by their grammatical behaviour (CN, NCN).

7.4 Singular Nouns, Plural Nouns

English plurals are normally fairly obvious and, like CNs themselves, are unmarked in the dictionaries of Volume 2. But there are exceptions:

moment of inertia	moments of inertia
angular momentum	angular momenta
energy quantum	energy quanta

In such cases, the relevant information is provided at the main entry in the Thesaurus, i.e. *moment*, *momentum*, *quantum*, or the equivalent in the TPD: *Moment*, *Impuls*, *Quant*. The dictionary symbol *pl* denotes the plural form in such cases.

The symbol *pl* also indicates so-called *plural nouns*, those which are not normally used in the singular, such as *nuclear forces* (Ge. *Kernbindungskräfte*). And the dictionaries contain terms broadly classified as *singular nouns*, those which may look plural at first sight (especially to German speakers) but in fact require singular verbs: *dynamics*, *nucleonics*, *physics*. Here the symbol *sg* is attached.

Wave mechanics *is* a field dealing with wave propagation.
Statics *is* an area of Newtonian mechanics.

A few nouns exhibit dual behaviour:

Ignition electrics *is* an interesting research topic.
The ignition electrics *are* due to be overhauled.

where normally two concepts are involved, for instance *ignition electrics* — the engineering field, and its plural counterpart — the electrical components, wiring and connections of an automobile ignition system.

7.5 Pair Nouns

The designation *pair noun* (*PN*) covers technical terms comparable to common expressions like *trousers, scissors, glasses* which behave like countable nouns in German (*Hose, Schere, Brille*) and require special attention in translation. Pair nouns take plural verbs regardless of whether the concept implied is singular or plural:

> Dividers *are* a tool used in Geometry, Map-Reading, Navigation.
> Special tongs *are* needed to handle the heated metallic frame.
> Pliers *are* necessary to twist the wire.

Just as the grammatical rules of general language lead to statements like:

> He put on *a pair of* trousers./… wore a new *pair of* glasses

rather than *"a trouser", *"a glass", so technical language too requires translators to think in terms of:

> *a pair of* dividers/… tongs/… pliers

for what German speakers regard as singular concepts (cf. *Spitzzirkel, Zange*).

7.6 Borderline Cases

The blanket category *non-countable noun* covers nouns with no plural form and no plural meaning. The verb is always singular:

> The bodywork of both vehicles *needs* some attention.
> The doping of the crystals *is* carried out individually.
> Lots of flux *has* to be used to get the solder to cover the areas.
>
> All machinery *is* to be switched off.
> Interference due to cosmic rays *upsets* this equipment.
> Attenuation of the waves received *is* easily compensated.

NCNs can co-occur with quantifiers *some/any/little/lots of/…* but not with the indefinite articles *a, an, one*, nor with numbers — *two, ten*, etc. They behave like the general expressions *bread, butter, information, progress.* But there are a number of borderline cases:

i. The difference between *singular* and *non-countable* nouns (sg/NCN) is that the former do not co-occur with *some, much*, etc. (*"some statics", *"much kinematics"), though the distinction is often due to *semantic* rather than *grammatical* reasoning, and may be marginal: *catalysis, electrolysis, induction.*
ii. Terms like *circuitry, interference, machinery* are perhaps a special type of *collective noun*, a consideration evident from the fact that their nearest German equivalents are *plural* expressions: *Schaltungen, Störungen, Maschinen.*

But this kind of hair-splitting is more for grammar enthusiasts than translators. Non-native English speakers merely need to realise that terms like *data, circuitry, electrolysis* take *singular* verbs and cannot co-occur with the article *a/an*, regardless of which of the two grammatical categories *sg/NCN* is designated in the Thesaurus.

7.7 Standard Grammatical Categories

A tiny minority of readers, those who have made a point of studying English grammar, may be disturbed by the author's usage of the expressions *pair noun, plural noun.* The following remarks should clear any misunderstandings:

i. The category *plural noun pl* conforms to what certain scholars would call *pluralia tantum*, namely nouns like *clothes, premises, remains, savings* or, using technical examples, terms like *nuclear binding forces, ignition electrics, soap suds.*
ii. *Pair nouns* PN, for instance *vernier calipers, scales, wire strippers*, conform to what grammarians regard as *summutation plurals*, both being special cases of *plural invariable nouns* (Quirk, Greenbaum et al.).

Categories, such as *collective nouns* — *audience, committee, staff, team*, do not occur in sufficient quantity in technical literature to warrant special attention. Terms like *machinery* are subsumed under the label of either non-countable or singular nouns (NCN, sg).

Rather than complicate the task of memorising and understanding dictionary symbols, the labels chosen sacrifice grammatical precision in favour of practical usefulness. They enable translators to interpret statements like:

> The *electrostatic charges* accumulate at the electrode and provide a total *electrostatic charge* of 20 micro-coulomb.

> The *total impedance* of the various *impedances* involved in the control circuit can be calculated using phasor diagrams.

which, though not necessarily intended as models for the translator to imitate, nonetheless occur frequently in engineering or scientific reports and, to the technologist, are perfectly normal.

Symbols differentiating noun classes provide yet another lexicographical device for improving the reader's dictionary interpretation and special translation skills. By distinguishing different *shades* of meaning, this facility enhances the reader's ability to interpret source material in English technical literature correctly. A scientist, industrial technologist or engineer does this passively and automatically.

7.8 Language Evolution

Most non-countable technical nouns are NCNs in general English too. The converse is not true, however, and translators who come across expressions like *energies, momenta, capacities, powers* in technical literature should not consider it substandard automatically on the basis of dictionaries or grammar textbooks. Sometimes technical language evolves in a slightly different way to general language. Translators who avoid or try to paraphrase such terms (*energy amounts, amounts of momentum*, etc.) purely as a result of their general-language instinct produce technical translations which look quaint and unrealistic. Those who go to the opposite extreme and fail to realise that **works, *heats, *inertias* are not acceptable make serious errors. Non-native speakers need to check literature rather than ordinary dictionaries to see whether an unfamiliar technical term, such as *apparatus*, is a CN or a NCN, and be perpetually on the lookout for false friends of the type *data* (NCN), Ge. *Datum/Daten* (CN).

Unit 8

Automotive Engineering

In contrast to some areas, the basic concepts of Automobile Technology are likely to be at least familiar to even the most "non-technically minded" linguists. Not all translators would recognise a *capacitor*, when they see one, but most people would know what an *engine* or *gearbox* looks like, and what they consist of. Automobile Engineering is a area where translation difficulties lie more in L2 term selection than in fundamental conceptual complexities. Technical descriptions in Chapter 8 are therefore brief, and primarily concern facilities for terminology acquisition, with the detailed description of engineering concepts taking a lesser role. Nevertheless, this chapter is probably the biggest of all, contains the largest number of microglossaries, the biggest individual thesaurus, and has links with virtually every other chapter.

So far the reader has been encouraged to study the book in the order in which it is presented on disk. This is still the case, as a substantial section of the terminology of Chapters 5–7 reappears in the field of Automobile Technology. Chapter 8 is taken out of sequence so that emphasis here can be placed more upon *linguistic* aspects of the technical discussion.

The contents of the chapter are as follows:

8.1 Preamble
- 8.1.1 Polyonymy
- 8.1.2 Associated Fields
- 8.1.3 Translator Training
- 8.1.4 Lexicology
- 8.1.5 Diachronic Change

8.2 Ignition, Fuel
- 8.2.1 Ignition Systems
- 8.2.2 Coil, Condenser, Capacity, Tension
- 8.2.3 Contact Breaker, Dwell Angle
- 8.2.4 Fuel Systems, Carburation

As the topic is comparatively simple, the chapter provides an opportunity of looking at technical terminology from a different perspective. Carefully chosen examples arranged in subsections separate from the main description of engineering concepts reveal a number of *general* features of specialised terminology: how meanings alter in the course of time, how misnomers appear and how archaic expressions outlawed in virtually every other branch of engineering can continue to persist when the original significance of the out-dated expression remains and the field is large enough to resist conformity. The phenomenon of *polyonymy* in technical literature, where a single concept has different designations in different parts of the English-speaking world, is introduced at the same time, and the chapter compares some approaches towards *translator training*.

8.1 Polyonymy

Automotive Engineering is not an area where differences between British and American usage can be ignored. Just as a British housewife is confused by the American usage of words like *muffin, biscuit, jelly* as opposed to *cake, scone, jam*, so a British automobile mechanic hesitates when an American customer complains of trouble with his *muffler* (Br. *silencer*). The industries grew up independently in the countries concerned. Since many of the concepts have become general everyday vocabulary, such as *engine/motor, boot/trunk, windscreen/windshield*, attempts to achieve conformity in the terminologies are repeatedly frustrated.

Figure 8A lists common concepts which are named differently by British and American automobile specialists. Apart from discrepancies in technical terminology, such as *antifreeze/defreezer, choke/air strangler, indicator/turn signal*, the list illustrates vocabulary contrasts relevant to motoring itself: *roundabout/turn circle, layby/rest area, traffic island/channelizing island*. Attention is drawn to major differences between British and American usage (e.g. *alternator/AC generator*) at various points in the chapter.

Spelling discrepancies — *tyre/tire, carburettor/carburetor, adaptor/adapter* appear in Figure 8A too, but trivial variations, such as *spark/sparking plug, bleed/bleeding nipple*), are not taken too seriously, owing to the fact that terminologies of different automobile manufacturers operating in the same country or even in the same city can vary due to internal company policy or to minor technical differences in the vehicles themselves.

8.2 Associated Fields

Automobile Engineering originated in the days of skilled craftsman as a by-product of what was then the *mechanical engineering* industry. Now, complex electronic devices are needed for tuning the engine, monitoring fuel consumption, or even operating the windscreen wipers. Chemical and metallurgical processes are required for *material-strengthening*, as well as rust-protection of the *bodywork, chassis* and *suspension system*. Sophisticated *robot technology* is employed for purposes such as *production assembly, welding, testing, cable-positioning*. Figures 8B–E provide microglossaries for the individual areas of the field, *ignition, carburation, braking, suspension*, and Figure 8F presents a lengthy microthesaurus.

Translators dealing with testing procedures for *automobile parts* must be careful to produce L2 material which is intelligible to native English speakers. This will not be the case if they overlook or confuse important parametric distinctions discussed in Chapters 1–2: *power/performance, torque/moment, stress/strain/tension, impulse/momentum, voltage/tension/emf, resistivity/resistance, capacity/capacitance*. Other relevant terminology appears in later chapters. Chapter 9 deals with Machine Technology (Ge. *Maschinenbau*) and takes a closer look at areas such as *engines, drive systems, bolts.* Chapter 13 concerns Construction Engineering (Ge. *Bautechnik*), but involves concepts like *stress, deformation, fatigue, creep* which relate to certain aspects of motor vehicles too: *vehicle suspension, steering joints, engine mountings.* Chapter 15 contains a section headed Auto-Electrics (15.4), which illustrates the conceptions underlying *alternators, dynamos, generators, starter motors* and *batteries* within the general framework of Electrical Engineering. And motor vehicles employ the full range of electronic devices (Chapters 6–7), especially in *ignition, fuel-injection, air-conditioning* and *lighting systems.*

8.3 Main Field

Chapter 8 discusses the main field of automotive engineering, which divides naturally into sections and subsections according to the engineering areas that have evolved around the motor vehicle itself: *engine, transmission, steering, suspension, cooling, ignition, fuel* and *brake systems.* Terminology is discussed either for its own sake, *cam, rocker shaft, valve tappet,* or to illustrate problems likely in translation: *pipe, tube, hose* (Ge. *Schlauch*); *connection, union* (Ge. *Anschluß*). A number of *technical verbs* occur in context, *choke, throttle, bleed,* encouraging the user to consult collocations in the TCD; terms belonging to special grammatical categories (NCN, PN, etc.) appear in context too: *paintwork, rubberwork, pliers.* The chapter is complemented by a selection of illustrations covering the main units or components of the various systems, *cylinder head, fuel pump, carburettor, wheel cylinder,* etc., and displaying their individual terminologies in both languages. Figures 8B–E present detailed microglossaries of the areas discussed.

In addition to describing the various component parts of the motor vehicle, the chapter draws attention to potential translation problems resulting from *misnomers, possible misinterpretations* and gradual *diachronic alterations* in meaning over the decades for which the automobile industry has existed. The reader's attention is also drawn to small variations in terminology among vehicle manufacturers, as well as to *jargon* expressions which have existed for many years and become fossilised within the field, *tension, condenser, water pump.* Rather than describe, for instance, *engine* or *brake parts* in detail, this unit highlights the additional features, with a small case study of three areas: *ignition, fuel* and *brake systems.*

8.4 Misnomers

The topic of *ignition systems* reveals examples of the surprising reluctancy of certain engineering disciplines to accept changes in terminology thrust upon them by other disciplines. Several generations have passed since the terms *low tension* and *high tension* (Ge. *Hoch-*, *Niederspannung*) were tantamount to normal technical English expressions, yet terms like *HT-lead*, *LT-lead*, *LT system* still linger in automobile literature like descendants of ancient dinosaurs. Moreover, the device *capacitor* is still frequently termed *condenser* in this field, and the misleading designation *capacity* from the viewpoint of the electrical or electronic engineer has not been entirely ousted by *capacitance*. But there is an even stranger misnomer.

For historical reasons, practical electronics enthusiasts employ the layman expression *coil* to denote the electrical device, more correctly designated as *inductor* (Chapter 2). An *inductor* consists of a *coil* of wire wound around an *iron core* and accessed via two *terminals*, *leads* or *electrodes*, mounted outside or on opposite sides of an *insulated casing* surrounding the device. A slightly more elaborate device involving two coils of wire (two *windings*) and with *four* external connections constitutes a *transformer*, one special case of which is the so-called *ignition coil*. The designation should be *ignition transformer*, yet this imminently more appropriate expression occurs nowhere in the literature. Thus, in a poor translation or a sloppily constructed English source text, the term *coil* can refer to three related but quite different concepts: *winding*, *inductor*, *transformer*. But what is even more amazing is that the German expression *Spule* exhibits the same parallel polysemy.

The extension of the basic meaning of *coil* to *inductor* is understandable, but there seems to be no reason why automobile technologists should use the same term to denote the concept *transformer*, in violation of every other field of Engineering. One can only hazard a guess as to the reason. In the nineteen-twenties, there was another type of ignition system, which is now completely obsolete (i.e. *magneto ignition*), where the demagnetisation of a large *coil* (now described as an *inductor*) created the ignition spark directly. If this is the true origin, the misnomer *ignition coil* may constitute one of the most ancient relics in the entire terminology of engineering.

8.5 Misinterpretations

Another area described in detail in the chapter is the *hydraulic brake system*, the system whereby *pressure* applied to the *footbrake pedal* in a motor vehicle is conveyed to the *roadwheels*. Provided that the brake system is *air-tight* and the presence of dust or grit is excluded, it should in theory require no maintenance at all. In practice, however, microscopic bubbles of air do eventually penetrate into the fluid after a few years. When this happens, or if the *fluid level* in the master cylinder begins to fall as the result of a leak somewhere, the fluid has to be drained and carefully replaced, a process termed *brake bleeding*.

Fluid is *bled* (Ge. *entlüften*) from each wheel cylinder in turn by attaching a device known as a *bleeder* (Ge. *Entlüftungsgerät*) to each of the *bleed valves* (cs: *bleed nipple*, *bleed screw*; Ge. *Entlüftungsschraube*) in turn and unscrewing the valves so that fluid runs out. The *bleeder* is merely a plastic bottle with a suitable tube which is then attached to the bleed nipple. The fluid level in the master cylinder must be continually *topped up* and the system *re-bled* until all air bubbles are removed. Air bubbles are compressible and adversely affect *braking efficiency*.

Terminology does not present a serious problem in this area, but unfortunately the expressions *bleeding* and *bleeder* coincidentally correspond to mild forms of abuse in Southern British colloquial English (both derive from *bloody*: *bloody hell*, etc.). In the spoken language, no confusion occurs because the *stresses* and *intonation patterns* in potentially ambiguous sentences are different. In a written text, however, translated sentences containing accidental ambiguities like:

i. If air enters the system you have to repeat the whole bleeding process.
ii. Attach the bleeding tube to the nipple as firmly as possible.
iii. Be careful not to upset the bleeder.

can appear either humorous or, to some people, mildly offensive.

As translations of normal engineering directives, there is nothing at all wrong with any of the three statements. This rare but interesting case reveals just how much translators need to keep their wits about them at all times.

8.6 Hierarchic Organisation

Unit 4 demonstrates how sequences of hierarchic term lists can be transformed into a single microthesaurus. Interested readers might care to experiment with the process in reverse, extrapolating large hierarchic term lists from the micro-thesaurus of Figure 8F. Section 8.1.4 illustrates the technique, showing how a structured bilingual microglossary of some 30 terms or more is obtained for the small field of *carburettors/fuel pumps.*

Hierarchic organisation provides valuable insight into the terminologies of small well-defined areas, as well as a basis for discussion of how particular terminologies, for instance particular automobile parts produced by specific manufacturers, differ from the general model. This system of organisation is very sensitive to global alterations in technical language, as well as to gradual alterations in the course of time: diachronic change.

8.7 Diachronic Change

Figure 8D contains a list of terms centred on the concept *drum brake.* If the book had been written in the nineteen-fifties this section of the glossary would have been virtually the same. *Drum brakes* have reached an evolutionary dead-end. No changes in their technology have taken place for many years; their terminology has become fixed and will probably remain so until drum brakes themselves are replaced by something else one day. Similar considerations apply to the *battery.*

Other areas are in a constant state of change. In the nineteen-eighties many cars still had two *cables* connected to the carburettor: the *accelerator cable* and the *choke cable.* Nowadays most vehicles have an automatic choke system. Only *accelerator cable* warrants inclusion in an up-to-date modern dictionary. Similarly, what was once an optional extra for more expensive vehicles, a *tandem brake system* (Ge. *Zweikreis-Bremsanlage*), which enables the *front brakes* to continue working when the *rear brakes* fail (or vice versa), has now become a standard fitting for all motor vehicles. There is no longer any need to distinguish between *single* and *tandem systems.* The terms are becoming obsolete and, like *choke cable,* no longer warrant inclusion in glossaries, except as a footnote.

And there are cases where technology itself seems to regress and expressions are rekindled which were thought to be vanishing. Throughout the history of automobile engines the *camschaft* which operates the *inlet* and *exhaust valves* was driven by a *chain,* the so-called *timing chain,* which never slipped, required no maintenance and often lasted the full lifetime of the *engine.* The nineteen-nineties witnessed a transition to *timing belts* operated by *timing pulleys* (Ge. *Riemenscheibe*), which slip or break perpetually and need to be changed at every 60,000 km service.

Language is in a constant state of change and technical language alters too in the course of time. One great advantage of the conceptual system of organisation is that obsolete terminology really stands out and can be removed or shifted, together with all subordinate terms associated with the obsolete concept, *as a complete block*. Updating a hierarchic glossary makes room for new concepts to flower with new associated terminology. Translators who acquire the intellectual expertise for handling conceptual arrangements and employ them for their own terminology not only keep engineering concepts alive in their own minds, they ensure that their data-bases never become dated.

8.8 Term Spotting

In view of the large volume of material covered, and the likelihood that the disk user will be vaguely familiar with the most frequent terminology, Chapter 8 devotes less space to the detailed description of concepts than its predecessors. The information is nevertheless there, but in concentrated form: within the glossaries. Consider the following extract from the microthesaurus of Figure 8F:

capacitance *Kapazität f*
m: capacitor; u: ignition system.
capacitor *Kondensator m*
u: electronic ignition system (capacitative-discharge ignition system).
capacity *Füllmenge f*
m: fuel tank; tu: litre; d: maximum amount of fuel which the tank can hold.
capacity *Kapazität f*
m: condenser; tu: microfarad; ps: capacitance.
capacity *Ladekapazität f*
m: car battery; tu: amp.hour (Ah).
condenser *Kondensator m*
u: conventional ignition system; cs: capacitor.
engine performance *Motorleistung*
d: how well the engine copes with fast driving, steep hills; ct: engine power.
engine power *Motorleistung*
m: engine; tu: kW; ct: former concept horsepower.
engine speed *Drehzahl f*
m: engine; tu: rpm (revolutions per minute); d: rotational speed.
pushrod *Stößel m*
d: rod attached to brake pedal operating the master cylinder plunger.
pushrod *Stößel m*
p: engine; d: rod lifted by a cam to push open one of the valves.

tappet rod *Stößel m*
cs: pushrod; a: camshaft, valve tappets.
throttle rod *Drosselklappenwelle f*
p: carburettor; d: rotatable shaft attached to the throttle.
tie-rod *Spurstange f*
d: rod connecting one of the front wheels to the steering assembly.

Note the following observations:

i. ***capacitance, capacitor, capacity***
There are at least three different interpretations of *capacity* within the field, distinguished among other things, by their *typical units* (*tu*): litre, microfarad, amp-hour (Ah). One interpretation is considered *substandard* by engineers outside the field. The *preferred synonym* (*ps*) is *capacitance*, a parameter related to two devices present in an *ignition system*, the *capacitor* and the *condenser*. The devices themselves are not radically different. One term is used in connection with *electronic ignition systems*, the other with *conventional* ones.

ii. ***power, performance, speed***
The German expression *Motorleistung* has two meanings. On the one hand it denotes *engine power*, a specific *parameter* measured in *kilowatts* (previously in the unit *horsepower*), on the other *engine performance*, a vague conception applicable more to the intuitive feeling of the driver of how well the engine *performs* on steep hills, motorways, etc. A related entry shows that *speed* is not translated by *Tempo* or *Geschwindigkeit* in this context, but *Drehzahl*.

iii. ***rod***
The terms *pushrod* and *tappet rod* are near synonyms in the context of *engines*, but there is a different concept *pushrod* in the area of *brake systems*, for which there is no alternative expression. By chance, German employs *Stößel* for both concepts, but not for *throttle rod*, *tie-rod*. Their German equivalents *Welle*, *Stange* reappear many times in the same glossary too, with other interpretations: *shaft*, *spindle*, *bar*.

Virtually every *chapter* of the disk ends with one or more detailed *thesauri*, each containing a multitude of examples like the above. The large, main *Technical Thesaurus* of Volume 2 includes many additional, often more complex engineering conceptions. Ideally, not only the *chapters* of the e-book should be read like a true book, from cover to cover, but also the thesauri. Translators who wish to improve their proficiency should conduct regular *term-spotting* or rather *polyseme-spotting* exercises of this nature.

Unit 9

Mechanical Engineering

Many of the younger generation of translators spend their lives glued to a monitor screen, clicking L2 substitutions from electronic dictionaries directly into their translations almost without bothering to read them. Speed is essential; quality is neglected. The reader is free to use (or rather *misuse*) the e-book in the same way, but valuable information is lost and, with a little prior preparation, there is no need for this sacrifice.

Now that the reader is becoming familiar with the *thesaurus technique*, it is convenient to take another giant leap forward, to Chapter 14, and examine other approaches to the disk. This time instead of presenting a *contents list* of the *chapter sections*, the unit begins with a *glossary*, Figure 14, the *Microthesaurus of Nautical, Aeronautical and Aerospace Engineering Terminology*:

aerofoil (–)
u: aircraft; p: body or wings; ex: fin, aileron, wing flap.

airframe *Flugzeuggerippe n*
d: structural unit of an aircraft or missile; ct: propulsion unit.

aileron *Querruder n*
p: movable control surface of an aircraft wing.

anticyclone *Hochdruckgebiet n*
u: meteorology; d: atmospheric region of high pressure; a: fine weather.

boosters *Trägerraketen pl*
d: lateral rockets which take a spacecraft into a terrestrial orbit.

buoyancy *Auftrieb m*
t: upward thrust; u: boat, ship or other floating body; ct: lift.

celestial body *Himmelskörper m*
ex: star, planet, moon, asteroid, comet.

communication satellite *Nachrichtensatellit m*
cs: comsat; u: telephone/radio links, broadcasting; ct: metsat.

cyclone *Tiefdruckgebiet n*
d: atmospheric region of low pressure; a: cloudy weather; ct: anticyclone.

drag *Widerstand m*

t: force opposing motion; ex: magnetic drag, frictional drag;
drag *Luftwiderstand m*
u: aircraft, balloon, helicopter; cs: atmospheric drag.
drag *Strömungswiderstand m*
u: boat, ship, missile, submarine; cs: current drag.
fairing *Verkleidung; Verschalung*
t: smooth structure intended to reduce drag; u: train, aircraft, etc.
fin *Flosse f*
u: aircraft; d: vertical aerofoil providing directional stability.
fuselage *Flugzeugrumpf m*
d: body of a plane; ct: hull (of a ship).
geostationary orbit *geostationäre Bahn*
d: orbit where satellite motion is synchronised to the earth's rotation.
ground control *Bodenstation f*
u: rocket launching, satellite control, space probe communication.
helm *Ruder n, Steuer n*
u: ship; d: position of control.
helm *Ruderspinne f*
d: wheel controlling the steering of a ship.
hull *Schiffsrumpf m*
d: body of a ship or submarine; ct: fuselage.
landing module *Landekapsel f*
p: spacecraft, space probe; d: section landing on a celestial body.
launch pad *Startrampe f*
u: rocket or missile launching.
lift *Auftrieb m*
t: upward thrust; u: aircraft, helicopter, balloon; ct: buoyancy.
meteorology *Wetterkunde f*
d: study of the earth's weather a: metsats (meteorological satellites).
oar *Ruder n*
u: rowing boat (Ge. *Ruderboot*); ct: rudder, helm.
orbit *Umlaufbahn f*
d: constant elliptical path of one celestial body around another.
pitch *neigen*
u: aircraft; d: turning about a lateral axis (e.g. nose down).
pitch *Neigung*
d: degree of turning about a lateral axis; ct: yaw, roll;
precipitation *Niederschlag m*
u: meteorology; d: rain, snow, hail, sleet.
roll *schlingern*
u: ship, aircraft; d: side to side movement; ct: pitch, yaw.
rudder *Ruder n*
u: ship; d: component at the stern used for controlling direction.
rudder *Ruder n*

u: aircraft; d: movable auxiliary aerofoil attached to the fin.
satellite *Satellit m*
t: man-made celestial body; ex: comsat, metsat, spysat.
satellite *Trabant m*
u: moon orbiting a planet, planet orbiting a star, etc.
space probe *Raumsonde f*
t: spacecraft or satellite for direct research on celestial bodies.
space shuttle *Raumfähre f*
t: reusable passenger stage of a spacecraft; a: space station.
space station *Raumstation f*
d: manned research satellite in a terrestrial orbit; ex: Mir.
stern *Heck n*
d: rear area of a ship, submarine or other sea-going vessel.
tail *Heck n*
p: aeroplane.
thrust *Schubkraft f, Schub m*
d: force providing motion in the required direction; tu: newton.
upward thrust *Auftrieb m*
t: force; a: material displacement; ct: weight; u: aircraft, ship, etc.
wing *Tragfläche f, Tragflügel m*
p: aircraft; ct: fuselage, tail, rudder, fin, aileron.
yaw *gieren*
u: aircraft, spacecraft; d: deviating from a straight course.

The glossary immediately reveals terms with several distinct conceptual interpretations, *drag*, *helm*, *pitch*, *rudder*, *thrust*, and likely pitfalls for inexperienced translators from German: *Auftrieb* (buoyancy, upward thrust, lift). The reader will find these and other related thesaurus entries in the *chapter*. They are discussed briefly in the extracts below.

The layout of the early disk chapters encourages the reader to first examine the text and later the glossaries. This unit tests the opposite approach as a viable alternative. It then extends this technique to relevant sections of another area of the book, Chapter 13: *Construction Engineering*.

Note: Chapter 14 is headed *Mechanical Engineering*, but really all it does is to tie up some loose ends from earlier chapters involving this field. The broad terminology of *mechanical engineering* occurs in Chapters 1, 8, 9, 12, 13 in other contexts. Chapter 14 deals with the *heavy* mechanical engineering aspects of *railway/railroad technology*, *nautical*, *aeronautical*, *aerospace engineering*.

9.1 Streamlining

Mechanical engineers design the *engines* of ships, submarines, aircraft and rockets; the *control systems* are the domain of electrical and electronics experts; the *structures* for assembling and stabilising large ships, heavy planes and multi-stage space rockets are the responsibility of construction engineers. *Nautical* and *aeronautical* engineers deal primarily with the design of *hydrodynamic hulls* (ships) and *aerodynamic air-frames* (aircraft). *Aerodynamic design* is important in the automobile industry too and certain other areas, such as railway vehicles. The term covering both *aero-* and *hydrodynamic design* is *streamlining.*

Nautical, aeronautical and aerospace engineers design sea-, air- or space-craft which move as efficiently as possible in the respective *medium.* Streamlined craft disturb the streams of air or water as little as possible and consequently create the minimum *turbulence* and minimum air/water resistance, so-called *drag.* The design of such craft requires extensive computer simulation and employs complex mathematical models. Only for equipment designed for use outside the earth's atmosphere, does streamlining lose its relevance, giving way to other crucial engineering factors: the effects of temperature extremes, high velocities, bombardment with ionised particles, etc.

The engines of planes, helicopters, rockets, submarines and ships develop *thrust* (Ge. *Schubkraft*), namely they expel accelerated particles which propel the craft forwards. The engines (or turbines) themselves and the type of fuel (steam, diesel, nuclear, chemical) differ greatly but the method of propulsion is the same in each case: Newton's Principle of *Action/Reaction.* The *force* with which particles are expelled creates an equal and opposite force, a so-called *reaction,* that propels the craft forwards. It functions even in vacuum (outer space).

In view of superficial similarities between Nautical and Aeronautical Technology from the viewpoint of the physicist or engineer, some German expressions (e.g. *Auftrieb, Heck, Ruder, Rumpf*) have different meanings within the two fields and radically different English equivalents. These terms are discussed in the sections below.

9.2 Drag, Lift, Thrust, Buoyancy

Four forces act upon a moving aircraft, such as a passenger plane: *thrust, drag, lift, gravity* (Ge. *Schub, Luftwiderstand, Auftrieb, Gewicht*). The force of gravity results from the *weight* of the plane and is counterbalanced by the *upward thrust* due to the weight of the atmosphere above this altitude, a force denoted by the term *lift*. The same forces act on a submerged submarine and the same terminology applies. *Thrust*, in other words the *forward* or *reverse thrust* developed by the engines of the plane/submarine, is opposed by *drag*, the friction forces resulting from the surrounding atmosphere/water. The forces acting on a ship or other *floating vessel* are the same, but as the vessel remains at the surface the terminology differs slightly: *thrust*, *drag*, *buoyancy*, *weight*. Certain *sea-going vessels*, such as hovercraft, move above the water surface. They depend on *lift* as well as *buoyancy* and are designed and operated by *aeronautical* rather than *nautical* engineers. German tends to employ the same expression *Auftriebskraft* for *lift*, *buoyancy* and *upward thrust*, and *Gewichtskraft* for both *gravity* and *weight*.

9.3 Fuselage, Hull, Helm, Rudder

Other terminology is common to both Nautical and Aeronautical Engineering though semantic connotations obviously differ. Boats have *rudders* to assist in steering. One of the *tail components* of a plane is also called a *rudder* and helps the plane to respond when entering or leaving a turn. The other tail components are: the *elevator*, which enables the plane to ascend or descend; the *horizontal/vertical stabilisers*, fixed attachments to the air frame. The *wing components*, the so-called *ailerons* and *flaps*, enable the plane to tilt to the right or left, a movement known as *banking*. Streamlining considerations for aircraft centre mainly on the angle and design of the *wing struts* in relation to the *fuselage*, and the facilities for raising the *landing gear*. For sea-going vessels, the shape of the *hull* determines the streamlining efficiency. Once again, analogies appear in German terminology: the same term *Rumpf* corresponds to *hull* or *fuselage*.

The German expression *Ruder* denotes both the *rudder* of a ship and that of an aircraft. It can also refer to *helm*, the place where the ship's direction is controlled, or, in another context, to the *oar* of a small boat or dinghy. *Helm* in English has a more basic meaning too. It refers to the *ship's wheel* (cs: *tiller*) which enables the captain or *helmsman* to steer the vessel. This well-known device with numerous spider-like handles corresponds to the German *Ruderspinne*. A third polyseme to mention in passing is the German *Heck*, corresponding to the *tail* of a plane or the *stern* of a sea-going vessel.

9.4 Pitch, Roll, Yaw

Just as locomotives are restricted to railways so planes moving in the same direction are channeled onto *airways*. These *flight lanes* are specified by altitude and help prevent collisions. For instance, certain eastbound flights are at 15,500 feet, westbound 1000 feet above them. Intercontinental flights take place mainly above 35,000 feet (10,000m) where use can be made of *jet streams*, fast moving air currents which help conserve fuel and shorten the flight time.

The pilot has three *control axes*: the *yaw*, a measure of the deviation from a straight course; the *pitch*, the amount by which the *nose* rises or falls with respect to the *tail*; the *roll*, the rocking movement of the wings. The same terms are used in connection with *rockets*, *missiles* and *spacecraft*, and *yaw* is used in *nautical navigation*. The expressions *yaw*, *pitch*, *roll* occur mainly as verbs (Ge. *gieren*, *neigen*, *schlingern*) and are equally common in Nautical and Aeronautical Engineering.

Note: *Pitch* occurs again and again in technology. It can imply: the distance between adjacent grooves of a *screwthread*; the distance between *adjacent atoms* in a stable solid; the frequency or acoustic level of a particular *note*, *sound* or *noise*; the *flying* and *sailing* terms described above. These are extensions of a common basic meaning but only remotely related. Hence, their German equivalents are entirely different: *Gewindeabstand*, *Atomabstand*, *Tonhöhe*, *Neigung*. As well as the polysemes *pitch*, an unrelated constructional-engineering homonym occurs too: *pitch*, a material used in road surfacing (Ge. *Pech*).

9.5 Construction Terminology

The approach demonstrated above employs examples taken directly from just a small extract of Chapter 14. The technique is also applicable to Chapter 13, which deals with *Construction Engineering*. The unit continues with a few extracts from this chapter, that are also relevant to *Mechanical* Engineering. This time the thesaurus (Figure 13) is not reproduced, but a lot of the information discussed in the rest of this unit can be deduced directly from it. The examples demonstrate another aspect of technical translation too: the fact that identical terminology occurs in different fields. Many of the terms reappear in earlier chapters and some have been mentioned already in this handbook.

The basic terminology of *construction design* derives from Physics and contains expressions like *tension*, *compression*, *stress*, *strain*, *moment*, *torque*, *torsion* described in Chapter 1. Forces applied to bridges, winches or crane jibs produce *stresses* in the materials concerned, which result in *strain*; different stresses, such as *tensile*, *bending* or *torsional stress*, cause materials to *fail* (to *fracture* or *rupture*) in different ways. Texts dealing with *material strengths* involve translators in terminologies derived from Metallurgy and Materials Science. Those concerning *structural design* require most of all a sound knowledge of *Solid-Body Mechanics*, the scientific basis of Construction Engineering.

9.6 Stress, Strain, Deformation

Chapter 1 discusses the broad engineering considerations of *bars*, *rods*, *girders*, *columns* and other *rigid members* used for structural applications, and it outlines the distinction *tie/strut*. This section returns to this topic, taking a closer look at the concepts *stress/strain* (Ge. *Spannung/Dehnung*).

The amount of *stress* occuring in any *member* of a construction, for instance a *bar*, depends on the position of the member relative to other members in the structure, as well as on the position and magnitude of the load. Simple structures involve two main varieties of stress: *tensile*, *compressive*. These lead to corresponding *strains*. *Tensile strain* occuring in a metal bar is often expressed as a percentage. It denotes the ratio of the minute increase in the length of the bar to its length when no load is applied and the bar is *unstressed*. In the same way, *compressive strain* refers to the proportional or percentage *reduction* in the length of a bar subjected to *compressive stress*, for example when the load is applied from above. The ratio of *stress* to *strain* is constant for the material concerned over a range of loading. The constant itself is termed *Young's Modulus of Elasticity* (symbol E) and has the same physical dimensions as stress.

Materials used in constructional applications, such as steel or aluminium, are said to exhibit *elastic* stress-strain behaviour, implying that deformations are temporary and that members made of such materials return to their original shapes immediately the stress is removed. Beyond a certain point however, the so-called *yield point* of the *stress–strain diagram*, permanent deformation (Ge. *bleibende Verformung*) does take place. The material stretches and *strain-hardening* is said to occur as the region of *plastic* stress-strain behaviour is entered. The above considerations concern *rigid members* (bars, girders, etc.) but are equally applicable to *wires*, *cable suspensions* and other non-metallic construction materials, such as *plastics* themselves (Ge. *Kunststoffe*).

9.7 Fatigue, Creep, Dislocation

Engineers dealing with composite structures have to analyse not just the *tensile* and *compressive* stress aspect but also the effects of *bending* and *torsion*. These lead to a three-dimensional system of stresses, the analysis of which requires complex mathematics involving the use of *stress vectors* to depict *linear* and *rotational forces* at incremental regions of the material. Combinations of these forces result in *shear stresses* which occur at an angle to the axis of the member concerned and can cause *material fracture* or *rupture* when the normal breaking strength (i.e. *ultimate rupture strength*) of the material is exceeded.

Perfectly elastic materials resume their shape when the stress is removed. *Perfectly plastic* ones adopt the new shape. *Ductile* materials can be permanently stretched, whereas *brittle* ones rupture immediately. Most materials exhibit all these properties somewhere along the *stress/strain curve*. When a material is subjected to *repeated stress*, this leads to *fatigue* (Ge. *Ermüdung*), a process whereby fractures occur at points of high stress (e.g. sharp corners, riveted holes) which then spread throughout the material.

The term *creep* (Ge. *Dehnung, Ausdehnung*) denotes the gradual increase in length or distortion in shape which occurs when a material is subjected to long periods of *constant stress*. This is a problem in the *turbine blades* of *jet aircraft*, especially due to the high temperatures involved. Soft metals (e.g. lead) and many plastics show considerable *creep* even at room temperature.

From the viewpoint of the materials scientist, *elastic deformation* involves an increase in *molecular separation* (the distances between molecules), whereas *plastic deformation* concerns whole rows of molecules slipping across one another to new locations. *Gaps* occur in the material leading to *dislocations* (Ge. *Störstelle*). Materials can be strengthened by introducing *foreign atoms* into the lattice which hinder the movement of dislocations, for example *carbon* diffused into *steel*. Materials without dislocations are very strong. But, so far, perfect metallic crystals have only been achieved for tiny pieces of material a few microns in length, so-called *whiskers*.

9.8 Gap, Hole, Foreign Atom, Impurity

The German terminology of this field, with expressions like *Lücke, Fremdatom* (E. *gap, foreign atom*), sometimes leads inexperienced translators and careless lexicographers to imagine they are dealing with the *same* terms in the field of

Semiconductor Materials (Chap. 5). But here the English equivalents are *hole, impurity atom.* The *purity standards* imposed on semiconductor materials are much more stringent than those to which metals are normally subjected; thus the science of construction materials employs expressions like *foreign atom, foreign substance, strengthening substance.* The contextual synonym *impurity,* when used, has very different connotations from the carefully injected individual atoms of phosphorus, arsenic, etc. employed in the semiconductor industry. Moreover, though the engineering concepts *Loch/Lücke* are not always as carefully distinguished in German, their English counterparts *hole/gap* have very different significances: *holes* imply *missing electrons,* whereas *gaps* are *missing atoms.*

Unit 10

Technical Polyseme Dictionary

By now, the reader has acquired a superficial command of the terminology described in six of the sixteen engineering chapters. It is time to introduce the dictionary sections of the second volume, the first of which is the *Technical Polyseme Dictionary* (TPD). This unit reorganises the *fields* covered by the dictionary, and discusses them collectively, so that their *codes*, the mnemonic labels attached to terminology, can be memorised in order of importance rather than at random. It then discusses the main features of the TPD itself.

The TPD supplies a basic engineering terminology for all linguists embarking on careers as German-English technical translators. It provides terminology in both languages corresponding to the key concepts discussed in Volume 1 and takes the reader beyond this initial basis to other fields of engineering. *Synonymy*, *hyponymy*, *contrast* and other semantic relations among engineering terms are revealed by a variety of organisational techniques involving *entry blocks*, *indentation*, *field codes* and *thesaurus descriptors*, and the dictionary provides information on *concept specification* and *polyseme recognition* for use in conjunction with the Technical Thesaurus. The TPD achieves a substantial increase in lexicological information compared to a conventional dictionary (an ordinary bilingual alphabetic glossary), with virtually no increase in size.

10.1 Subject Fields

The set of field codes below are the most frequent ones. They denote global engineering fields and correspond to major sections of Volume 1.

AUTO	Automobile Technology	AERO	Aeronautics
CHEM	Chemical Engineering	HYD	Hydraulic Engineering
CONS	Construction Engineering	NAUT	Nautical Engineering

DPS	Data-Processing Systems	POW	Power Transmission
ELEC	Electrical Engineering	RAIL	Railway Engineering
ELNC	Electronics	ROCK	Rockets & Missiles
EMAT	Engineering Materials	STEA	Steam Engines
MACH	Machine Technology	WAST	Waste Disposal
MECH	Mechanical Engineering		
NUCL	Nuclear Technology		
SEMI	Semiconductor Technology		

Terms on the left cover the contents of one or more entire engineering chapters. Those on the right, e.g. *Hydraulic Engineering* (supply systems, waste-water flow), *Steam Engines* (locomotives, steamships), are associated with broader areas, such as *Mechanical Engineering*, but have specific terminologies of their own. The dictionaries also contain names of subject fields themselves, such as *Chemie*, *Dynamik*, *Schallehre* (Chemistry, Dynamics, Acoustics). These are entered with the field code: SUBJ (Subject Field).

The book employs the following field codes for Physics branches:

ACU	Acoustics	MAGN	Magnetism
ASTR	Astronomy	MECH	Mechanics
ELEC	Electricity	OPT	Optics
ELSC	Electrostatics	OSCN	Oscillations
FLUD	Fluid Dynamics	WAVE	Waves
GAS	Gases and Vapours		

Observant readers will notice that the codes MECH, ELEC (Mechanics, Electricity) reappear but designate apparently different fields to those listed above (Mechanical, Electrical Engineering). This is because the engineering fields stem directly from the Physics branches and employ all the terminology of the latter. It is true that a physicist would not be familiar with the complete terminologies of Mechanical and Electrical Engineering, but the difficulties of differentiating Mechanics and Electricity from their engineering counterparts, and the absence of any real advantage to the translator in doing so, mean that the same dictionary symbol is employed for both the engineering field and its smaller scientific subfield. There are Physics terms which appear throughout the subject, and throughout engineering (e.g. *energy*, *power*, *work*), but are not restricted to any of the branches above. These are labelled PHYS (Physics).

Just as there is no clear dividing line between certain engineering fields and their original scientific basis in Physics, so terms from Mathematics and other areas of Science occur in translation assignments too. The book provides terminology from the following areas:

ATOM	Atomic Physics	MEAS	Precision Measurement
CHEM	Chemistry	MET	Meters & Gauges
GEOM	Geometry	NUCL	Nucleonics
GRAF	Graphs & Charts	PHOT	Photography
LAB	Laboratory Apparatus	RADN	Radiation
MATH	Mathematics	SUBS	Chemical Substances
MATS	Materials Science		

Two small remarks are appropriate:

i. The fields *Geometry* (GEOM) and *Graphs* (GRAF) are distinguished from their main field *Mathematics* (MATH) in the hope that, in the dictionaries, translators may recognise certain concepts from their early schooldays. Terminology from areas of Mathematics less familiar to linguists, such as *Matrix Algebra, Differential Calculus, Vector Analysis, Trigonometry* is not distinguished in this manner.

ii. The codes CHEM and NUCL denote the scientific fields *Chemistry* and *Nucleonics*. They reappear earlier in the section, where they designate *Chemical/Nuclear Engineering*. The situation is analogous to that of MECH/ELEC above.

The remaining codes correspond to small subfields and related fields subsidiary to the mainstream of engineering. They happen to have fairly large terminologies.

1. Automobile Engineering (AUTO)

BRAK	Braking Systems	IGN	Ignition Systems
ENGN	Engines	RUNN	Running Gear
FUEL	Fuel Systems		

2. Electrical and Electronic Engineering (ELEC/ELNC)

BATT	Batteries	SDEV	Semiconductor Devices
EENT	Electronic Entertainment	TRAN	Transformers
LAMP	Lamps & Fittings	TV	Televisions, Monitors
REM	Remote Control Systems		

3. General Equipment (GEN)

BIKE	Bicycles	HOUS	Household
CLOK	Clocks & Watches	OFF	Office Equipment
FRIG	Refrigerators, Freezers	TOOL	Tools

Though it would be possible in some cases to replace the field code by a broader subject designation (e.g. AUTO, ELEC, MECH), important information might be lost. Moreover, it is often difficult to decide whether for instance the terminology of *Batteries* (BATT) belongs under *Electrical* or *Chemical Engineering*, and whether *Clocks & Watches* (CLOK) is part of *Mechanical Engineering*. The codes remain, but as the terminology is needed less frequently these codes are the last ones the reader needs to memorise.

10.2 Variation, Gender

The TPD adopts the following convention regarding the two main variants of technical English: expressions used only in British English, but not in American, are marked *Br*; those appearing in American but not in British are designated *Am*.

There are not many engineering areas where discrepancies occur and even where they do exist (automobiles, railways, household plumbing) subject specialists on both sides of the Atlantic are generally familiar with both variants. To enable the translator to achieve consistency, however, especially in work involving technical advertising, the above symbols are also used when a particular expression is the *preferred* form in either one of the two variants.

The usual *gender* symbols are employed, but as unnecessary genders occupy valuable space, the TPD adopts simplifications. These are defined explicitly in the dictionary introduction and may be evident in the examples following.

10.3 Polysemy

Entries derived from polysemous root terms such as *Fläche* (area, surface, space, face, interface) are organised as follows:

Fläche *f*:

Fläche	*MATH*	area
Fläche	*PHYS*	surface
bearbeitete Fläche	*MACH*	machined surface
Bodenfläche	*HOUS*	floor space
Grenzfläche	*SDEV*	junction interface
Kontaktfläche	*IGN*	contact face
Kristallfläche	*MATS*	crystal face

Oberfläche	*GEN*	surface
Querschnittsfläche	*MATH*	cross-sectional area
Reibungsfläche	*MECH*	friction surface
Flächeninhalt *m*	*MATH*	surface area

The main entry is complete with gender but without an English translation of the lexeme itself. Compound expressions are indented and appear as a *block* in alphabetic order, without duplication of gender. Basic interpretations of the polyseme (e.g. *Fläche — area, Fläche — surface*) precede the set of compounds and may modify them in specific contexts, for example *surface area, floor area, machined area* are unlisted possible interpretations of *Fläche, Bodenfläche, bearbeitete Fläche.* Terms containing the main entry in initial position (e.g. *Flächeninhalt*) follow the main entry block, provided that they are adjacent alphabetically.

10.4 Hyponymy

Not all entry blocks concern polysemes. Some involve hyponyms which are grouped together for convenience and appear as follows:

Bewegung *f*	*PHYS*	motion, movement
Drehbewegung	*ASTR*	rotary motion
gleichförmige Bewegung	*PHYS*	uniform motion
Schwingbewegung	*ACU*	oscillatory motion
Wärmebewegung	*ATOM*	thermal motion

Hyponym entry blocks are easily distinguishable from those containing polysemes, as the main entry or common root of the hyponyms (e.g. *Bewegung*) is provided with an English equivalent itself. Any alternatives cover the compounds too. Thus *Drehbewegung* is listed as *rotary motion*, but *rotary movement* can be inferred as an non-preferable alternative from the structure of the group.

10.5 Homonymy

Obvious *homonyms* such as *Scheibe* (disc, pulley, washer) as opposed to *Scheibe* (window, windscreen, pane) are listed as separate entry blocks in order to distinguish the two sets of meanings:

Scheibe *f*:		
Bremsscheibe	*BRAK*	brake disc
Mattscheibe	*PHOT*	focussing screen
Riemenscheibe	*MECH*	pulley
Schleifscheibe	*MACH*	grind wheel
Unterlegscheibe	*MECH*	plain washer
Scheibe *f*(2):		
Fensterscheibe	*HOUS*	window pane
getönte Scheibe	*AUTO*	tinted window
Seitenscheibe	*AUTO*	side window
Windschutzscheibe	*AUTO*	windscreen (Am. windschield)

Other examples are: *Welle* (shaft, spindle), *Welle* (wave, waveform); *Stärke* (strength, intensity), *Stärke* (starch); *Lehre* (gauge, meter), *Lehre* (e.g. *Gruppenlehre* — group theory). A few homonyms are distinguished by gender, such as *Messer,-m* (gauge, meter), *Messer,-n* (knife, cutting tool) or *Leiter,-m* (conductor, waveguide), *Leiter,-f* (ladder). These too are listed separately.

10.6 Entry Blocks

The above approach, namely the splitting of lexically similar terminology into separate entry blocks is also employed for *related* expressions (i.e. polysemes as opposed to homonyms) when the number of compounds happens to be very large. For example, those entered at *Stoff* (substance), *Stoff*[2] (material):

Stoff *m*:		
Stoff (cs: Substanz)	*CHEM*	substance
gefährlicher Stoff	*CHEM*	hazardous substance
giftiger Stoff	*CHEM*	toxic substance
gelöster Stoff	*CHEM*	solute
Schadstoff	*CHEM*	harmful substance
strahlender Stoff	*RADN*	radio-substance
Trägerstoff	*CHEM*	substrate
Stoff(2):		
Stoff (cs: Werkstoff)	*GEN*	material
Baustoff	*CONS*	building material
Dotierstoff	*SEMI*	dopant

Halbleiterwerkstoff	*SEMI*	semiconductor material
Isolierstoff	*ELEC*	insulation
Klebstoff	*MACH*	adhesive
Kunststoff	*MAT*	plastic
Rohstoff	*GEN*	raw material

The structures of the blocks concerned enable the reader to infer the correct superordinate engineering concept, for instance that *gelöster Stoff* (solute) concerns a "*substance*", whereas for *Dotierstoff* (dopant) the global term is "*material*". Double entry blocks for polysemes are distinguished from those for homonyms by the fact that the gender is not repeated in the second or subsequent blocks. Moreover, the first term may be accompanied by a small thesaurus statement which helps to narrow the global concept and identify the criteria for the blocks. The statements "cs: Werkstoff" and "cs: Substanz" reveal that the German expressions *Werkstoff*, *Substanz* are *contextual synonyms* of *Stoff* in the areas indicated.

After dividing terminology the blocks may still be large. Indeed, the above examples are merely *extracts* from the TPD at the entry *Stoff*. The advantages of further subdivision must be weighed against the accessibility. Thus the wide variety of interpretations of the German term *Kraft* (force, lift, thrust, tension, power, etc.) are presented according to the following structure:

Kraft *f*:		
Kraft	*PHYS*	force
Abstoßungskraft	*ELSC*	repulsive force
Antriebskraft	*MECH*	propulsive force
…		
Gewichtskraft	*PHYS*	(force of) gravity
Hebelkraft	*PHYS*	leverage
Kernbindungskräfte	*NUCL*	nuclear (binding) forces
…		
Reibungskraft	*PHYS*	friction
Saitenspannkraft	*ACU*	string tension
Schubkraft	*AERO*	thrust
Stoßkraft	*PHYS*	force of impact
Verbiegungskräfte	*MECH*	bending forces
Kraft$^{(2)}$:		
Bremskraft	*BRAK*	braking power
Kernkraft	*NUCL*	nuclear power

Spannkraft	*MAT*	elasticity
Wärmekraft	*POW*	heat

There is a large block of alphabetic terms, all of which concern the physical quantity *force* (Chapter 1), followed by a small block of mixed terms none of which have anything to do with the concept *force*. The entries below *Kraft*$^{(1)}$ are related hyponyms stemming from the basic engineering concept *force*, whereas those grouped under *Kraft*$^{(2)}$ involve unrelated polysemes from entirely different technical fields.

Such decisions may seem arbitrary at first, but as the reader progresses and begins to understand the content of Volume 1 the relevant lexicographical criteria become apparent, by close comparison of the entries themselves.

10.7 Indentation

It is conceivable that terminology relating to the concepts *motor/engine* (Ge. *Motor*) could be arranged in the TPD as follows:

Elektromotor	*ELEC*	motor, electric motor
Drehstrommotor	*ELEC*	three-phase motor
Gleichstrommotor	*ELEC*	dc motor
Hilfsmotor	*AUTO*	auxiliary motor
Hilfsmotor	*ELNC*	servomotor
Synchronmotor	*ELEC*	synchronous motor
Wechselstrommotor	*ELEC*	ac motor
…		
Motor	*AUTO*	engine; motor (Am.)
Dieselmotor	*AUTO*	diesel engine
leistungsstarker Motor	*AUTO*	high-performance engine
Magermotor	*AUTO*	lean engine, lean runner
schadstoffarmer Motor	*AUTO*	clean-exhaust engine
Ottomotor	*AUTO*	petrol engine

The separate entry blocks reveal that for the terms listed below *Motor* (*engine*) the substitution *motor* is common in American (*diesel motor*, etc.), whereas for expressions below *Elektromotor* the substitution *engine* is not possible under any circumstances. "*Three-phase engine*", "*dc engine*", "*servoengine*" … are serious translation errors.

Nonetheless, if a separate entry block is incorporated for *Elektromotor* at a different point in the dictionary (where words begin *Elek-* rather than *Motor-*) the user needs to know that for instance *Drehstrommotor* is to be found there and not at *Motor*. The lexicological dilemma is resolved by indenting the blocks and placing them together as a joint entry "*Motor*".

Motor *m*:		
Motor	*AUTO*	engine; motor (Am.)
Dieselmotor	*AUTO*	diesel engine
leistungsstarker Motor	*AUTO*	high-performance engine
Magermotor	*AUTO*	lean engine, lean runner
schadstoffarmer Motor	*AUTO*	clean-exhaust engine
Ottomotor	*AUTO*	petrol engine
Elektromotor	*ELEC*	motor, electric motor
Drehstrommotor	*ELEC*	three-phase motor
Gleichstrommotor	*ELEC*	dc motor
Hilfsmotor	*AUTO*	auxiliary motor
Hilfsmotor	*ELNC*	servomotor
Synchronmotor	*ELEC*	synchronous motor
Wechselstrommotor	*ELEC*	ac motor

Indentation provides a convenient alternative to entry block separation when important semantic features are not evident from the terms themselves. Here for instance, the features of *Elektromotor* relate to *Synchronmotor* but not to *Magermotor*.

10.8 Secondary Indentation

When the number of compounds belonging to a certain block is not too great, indentation provides a way of separating different lexicological aspects of the terminology, especially *polysemy* from *hyponymy*. For example:

Träger *m*:		
Dachgepäckträger	*AUTO*	roof rack
Eisenträger	*CONS*	steel girder
Flugzeugträger	*NAUT*	aircraft carrier
Ladungsträger	*SEMI*	charge carrier
frei beweglicher L.	*MATS*	mobile charge carrier

Majoritätsträger	*SEMI*	majority carrier
Minoritätsträger	*SEMI*	minority carrier
Nadelträger	*STER*	stylus holder
Radträger	*AUTO*	hub carrier

Here, the English translation of the hyponym *Majoritätsträger* (majority carrier) is located below the polysemous entry *Träger* (unspecified), indented beneath its generic superordinate *Ladungsträger* (charge carrier). Its antonym *Minoritätsträger* (minority carrier) is also evident from the structure of the entry block, and so is a related concept: *frei beweglicher Ladungsträger* (mobile carrier).

Consider another example:

Bahn *f*:

Bahn	*PHYS*	path
Flugbahn	*ROCK*	trajectory
Hyperbelbahn	*NUCL*	hyperbolic path
Kreisbahn	*NUCL*	circular path
Kurvenbahn	*ROCK*	curved path
Spiralbahn	*NUCL*	spiral path
Eisenbahn	*RAIL*	railway, railroad (Am.)
Fahrbahn	*AUTO*	lane
Führungsbahn	*MECH*	guide track
Straßenbahn	*GEN*	tram, streetcar (Am.)
Umlaufbahn	*PHYS*	orbit
Ellipsenbahn	*ASTR*	elliptical orbit
Kreisbahn	*ASTR*	circular orbit

The example demonstrates that the English term *trajectory* is a hyponym of *path*, and that the German expressions *Ellipsenbahn*, *Kreisbahn* are possible hyponyms of *Umlaufbahn* (orbit). The hyponym information is separate from that relating to the polyseme *Bahn* (railway, lane, track, etc.).

Compromises are necessary though, as for the separate entry blocks of the previous sections. The advantages of secondary indentation must be weighed against the increased access time. Moreover, the limited space of the printed page or monitor screen makes further indentation difficult. Decisions such as whether to place the sub-block *Umlaufbahn* below *Bahn* and thus indicate that *orbit* is a hyponym of *path* depend not on semantic criteria but merely on the number of letters in the terms involved.

Entry block divisions and indentations provide valuable intuitive insights into polysemy, homonymy, hyponymy, synonymy, antonymy and other lexicological aspects of technical terminology. Terms mentioned in Volume 1 often reappear in compounds throughout the TPD where the structural arrangements reinforce the reader's understanding of engineering concepts once again. In a few cases, however, structural organisation alone does not specify terminology adequately enough for the purposes of translation. The remaining sections indicate how such problems are overcome by combining structural arrangements with thesaurus definitions.

10.9 Concept Specification, Target Language

For a word like *Wechselstromwiderstand* the given field code (i.e. ELEC) and the English translation (*impedance*) are not sufficient to distinguish it from *Scheinwiderstand*, which is listed with the same field code and the same English equivalent, even though the German terms are not interchangeable. Rather than ignore such problems, the Polyseme Dictionary attempts to resolve them. It does so, by using thesaurus descriptors. The entries mentioned are differentiated as follows:

Wechselstromwiderstand	*ELEC*	impedance (t: device)
Scheinwiderstand	*ELEC*	impedance (t: parameter)

which indicates that one term concerns an electrical device, the other a measurable parameter of such devices. The technique can be extended for small indented blocks. If space is limited the definition may appear on the left:

Leistung (u: power generation)	*ELEC*	volt-amperage
Blindleistung	*ELEC*	reactive volt-amperage
Scheinleistung	*ELEC*	volt-amperage
Wirkleistung	*ELEC*	effective power
Leistung (u: circuit design)	*ELNC*	power
Blindleistung	*ELNC*	reactive power
Scheinleistung	*ELNC*	apparent power
Wirkleistung	*ELNC*	power, real power

This entry reveals that the terms *Blind-*, *Schein-*, *Wirkleistung* in the heavy electrical engineering context of *power generation* are translated quite differently to their counterparts in the distantly related electronics field of *circuit design*. Different terminology has arisen for slightly different concepts in one language but not in the other.

This method of dictionary organisation is a powerful one for technical polysemes which are easily overlooked. Consider the Mechanical Engineering concept *Antrieb*:

Antrieb *m*:		
Antrieb (u: motor vehicle)	MECH	drive, drive system
Allradantrieb	AUTO	four-wheel drive
Heckantrieb	AUTO	rear-wheel drive
Vorderradantrieb	AUTO	front-wheel drive
Antrieb (u: force transfer)	MECH	drive, drive mechanism
Kettenantrieb	MECH	chain drive
Riemenantrieb	MECH	belt drive
Zahnradantrieb	MECH	gear drive
Antrieb (u: plane, ship)	PHYS	propulsion system
Dampfantrieb	NAUT	steam propulsion
Düsenantrieb	AERO	jet propulsion
Raketenantrieb	ROCK	rocket propulsion
Antrieb (tu: newton)	PHYS	propulsive force
Antrieb (tu: newton)	MECH	driving force, drive

By avoiding the temptation to list the various compounds alphabetically this entry provides clues to the semantic distinction between *drive* and *propulsion*: one term is used in connection with *motor vehicles*, the other is associated more with *planes, ships, rockets*. Moreover, it differentiates the meanings *drive system* and *drive mechanism*, and reveals that *four-wheel, front-wheel, rear-wheel drive* are hyponyms of *drive* in reference to *automobiles* whereas the hyponyms *chain, belt, gear drive* denote alternative *mechanisms* for the transfer of a *driving force*. The fundamental meanings of *Antrieb*, both as a physical quantity (i.e. *propulsive force*, typical unit: *newton*) and as a parameter in Mechanical Engineering calculations (*driving force*), are included separately and the reader might deduce that *propulsion* and *drive* are possible substitutions here too.

10.10 Concept Specification, Source Language

The Thesaurus technique is applicable to the German entry terms as well as the English translations. Thus the two entry blocks appearing at *Bohrer* are differentiated by:

Bohrer (p: Bohrmaschine)	*MACH*	drill bit
Bohrer (t: Bohrmaschine)	*MACH*	drill

indicating that one meaning is *part of* a drill (Ge. *Bohrmaschine*), the other *a type of* drill itself. Another example is the set of acids entered under *Säure*, for instance:

Ameisensäure (cs: Methansaüre)	*CHEM*	formic acid
Äthansäure	*CHEM*	ethanoic acid
Carbonsäure (ex: CH_3COOH)	*CHEM*	carbon-based acid
Essigsäure (cs: Äthansäure)	*CHEM*	acetic acid, vinegar
Kohlensäure (d: HCO_3)	*CHEM*	carbonic acid
Methansäure	*CHEM*	methanoic acid

These entries reveal the additional information that *Methansäure*, *Äthansäure* are contextual synonyms (cs) of *Ameisensäure*, *Essigsäure*, thus providing links to the entries *methanoic*, *ethanoic acid*; the distinction between *Carbonsäure* and *Kohlensäure* is that one term designates a class of acids (for example CH_3COOH), the other a particular acid (HCO_3).

Entries in the TPD are arranged into subgroups when the advantage in revealing polysemy, homonymy, hyponymy, synonymy or contrast outweighs the lexicographical advantage of a simple alphabetic arrangement. Large distinct subgroups tend to become separate dictionary entries regardless of the attribute responsible for the size of the group. The reader recognises polysemes immediately, by virtue of the field codes or the dictionary structure itself, thereby avoiding elementary *mistakes* in translation even if the desired compound is not in the dictionary. Thus the TPD is not intended as a global dictionary of *all* engineering conceptions. It is a didactic model involving *samples* of basic terminology used repeatedly throughout engineering.

Unit 11

Chemical Engineering

The approach to the terminology of an engineering discipline varies according to the field itself. Electrical Engineering involves difficult concepts, *reactance*, *phasor*, *commutator*, etc., which can only be described in context, while Mechanical Engineering concerns entities that are easier to visualise, so that often translators merely need clearly constructed glossaries and large thesauri of the type given (Figures 8A–F, etc.). Chemical Engineering (Chapter 10) occupies a midway position. The Contents List indicates a different approach too:

10.1 Metal, Non-Metal
- 10.1.1 Inert Element, Noble Gas
- 10.1.2 Precious Metal, Noble Metal, Base Metal
- 10.1.3 Light Metal, Heavy Metal

10.2 Solid, Liquid, Gas, Vapour
- 10.2.1 Allotrope, Polymorphic Substance
- 10.2.2 Fluid, Emulsion, Colloid, Gel
- 10.2.3 Solute, Solvent, Solution
- 10.2.4 Bond, Radical

10.3 Acid, Alkali, Base
- 10.3.1 Halogen, Salt
- 10.3.2 Acidity, Basicity, Alkalinity

10.4 Oxidation, Reduction, Anion, Cation
- 10.4.1 Reagent, Catalyst, Inhibitor
- 10.4.2 Endothermic/Exothermic Reaction

10.5 Isomer, Polymer, Hydrocarbon, Carbohydrate
- 10.5.1 Plastics, Polymers
- 10.5.2 Alkane, Alkene, Alkyne
- 10.5.3 Aliphatic/Aromatic Compounds
- 10.5.4 Sucrose, Glucose, Fructose
- 10.5.5 Inorganic Compounds

The chapter is divided into sections whose headings relate, almost entirely, to *contrasting* terminology: *precious/base metal*, *solute/solvent*, *anion/cation*, *catalyst/inhibitor*, *aliphatic/aromatic compound*. There are several glossaries and a lengthy *Microthesaurus of Chemical Terminology*.

This field has a device for distinguishing related concepts, as powerful as the technique of Chapters 1–2 for differentiating *parameters*, namely the *chemical formula*. This is demonstrated by an extract from Figure 10A, that also illustrates another lexicological arrangement, the reverse-sorted alphabetic dictionary, or simply *reverse dictionary*:

carbonic acid	H_2CO_3	Kohlensäure
nitric acid	HNO_3	Saltpetersäure
phosphoric acid	H_3PO_4	Phosphorsäure
chloric acid	$HClO_3$	Chlorsäure
hydrochloric acid	HCl	Salzsäure
sulphuric acid	H_2SO_4	Schwefelsäure
nitrous acid	HNO_2	saltpetrige Säure
sulphurous acid	H_2SO_3	schweflige Säure
calcium carbide	CaC_2	Calciumcarbid
zinc iodide	ZnJ_2	Zinkjodid
hydrogen sulphide	H_2S	Schwefelwasserstoff
carbon disulphide	CS_2	Schwefelkohlenstoff
sodium chloride	$NaCl$	Natriumchlorid
hydrogen chloride	HCl	Chlorwasserstoff
sodium fluoride	NaF	Natriumfluorid
hydrogen fluoride	HF	Fluorwasserstoff
carbon dioxide	CO_2	Kohlendioxid
sulphur dioxide	SO_2	Schwefeldioxid
sodium hydroxide	$NaOH$	Natriumhydroxid
hydrogen peroxide	H_2O_2	Wasserstoffperoxid
potassium sulphate	K_2SO_4	Kaliumsulfat
calcium phosphate	$Ca_3(PO_4)_2$	Calciumphosphat

potassium permanganate	$KMnO_4$	Kaliumpermanganat
calcium carbonate	$CaCO_3$	Kalziumcarbonat
sodium chlorate	$NaClO_3$	Natriumchlorat
potassium nitrate	KNO_3	Kaliumnitrat
sodium sulphite	Na_2SO_3	Natriumsulfit
potassium nitrite	KNO_2	Kaliumnitrit

The glossaries do not cover a wide area, but there is a lot of terminology. As so often in technical literature, contrasts occur which create problems for translators. The sections below examine some of these contrasts and discuss recent terminological alterations.

11.1 Chemical Terminology

Although new *materials* are devised every year, especially in the *polymer* and *plastics* industries, the basic terminology of chemical engineering remained reasonably constant, until recently. In the nineteen-seventies, chemists were using non-systematic but long-established expressions like *"nitrous oxide", *"nitric oxide", to denote gases with chemical formulas NO, N_2O, without feeling bothered by the denotational contradiction with substances like *carbon monoxide* (CO). This situation has changed. The gases NO, N_2O, NO_2 now have the more respectable designations *nitrogen monoxide, dinitrogen monoxide, nitrogen dioxide*, and the obsolete expressions *nitric/nitrous oxide* (still found in some dictionaries and antiquated data bases) are free to acquire new significances. Some young chemists already use these expressions as synonyms for the collective term covering the entire group of gaseous substances whose molecules consist entirely of nitrogen and oxygen, the *nitrogen oxides* (Ge. *Stickoxide).* Their general chemical formula NO_x has led to a third colloquial synonym: *noxies.*

New labels have appeared for other laboratory substances too, for instance *ethanoic acid* (acetic acid), *tri-oxygen* (ozone), *sodium hydrogensulphate* (sodium bisulphate). The new names correspond better to the respective *chemical compositions* than the old ones. The new terminology is settling down in the English-speaking world, and the German equivalents will probably be adjusted accordingly.

The situation is similar to that of the early semiconductor electronics industry when terms like *"condenser", *"capacity", *"cycles per second (cps)" were replaced by the more appropriate expressions *capacitor, capacitance, hertz (Hz).*

German did not respond as rapidly then and still uses the original, established terminology *Kondensator, Kapazität.* But the degree of international collaboration and cooperation in the chemical industry today is much better than that of the early, highly competitive electronics industry. The real impetus towards standardisation of chemical nomenclature, however, has come about as a result of the vast numbers of new substances and materials, discovered or designed by the chemical industries each year. These require appropriate names, in a well organised systematic terminological framework, to prevent confusion with other unrelated substances. Just as biologists rename different species of animals according to where they fit within the conceptual classification systems, or paleontologists reclassify early homonids, so chemists are renaming their substances and materials too. Translators need to be aware of this problem and the extent to which the new terminology encroaches upon industries outside chemical engineering itself.

The rest of the unit selects a small sample of terminology from the chapter to clarify some of the conceptions mentioned.

11.2 Metal, Non-Metal, Inert Element

The distinction between *metallic* and *non-metallic elements*, so-called *metals* and *non-metals*, is made by drawing a diagonal line through the *Periodic Table* (Figure 3A). All elements above or on the line joining the elements *boron* to *astatine* are classed by the chemist as *non-metals.* The rest are *metals.* Thus *carbon, chlorine, sulphur* are regarded as non-metals, whereas *aluminium, copper, tin* are metals. Elements of other substances less familiar in everyday life are classed as metals too, *antimony, cadmium, germanium, manganese,* as are indeed the overall majority of elements, including two which are rarely (if ever) solid: *mercury, hydrogen.* The chemical expression *metal* co-exists alongside the normal engineering conception *metal,* that mostly refers to *alloys* of metallic elements (Ge. *Legierung*), terms like: *iron, copper, bronze, zinc-plate, sheet metal.*

Metals are distinguished by their densities. Those below $5 g/cm^3$ (e.g. magnesium, aluminium) are referred to by chemists as *light metals*; those above this value (iron, copper, lead, gold, cadmium) are *heavy metals.* As certain heavy metals tend to remain in the human body with harmful consequences (especially lead and cadmium), the expression *heavy metal* sometimes has the negative connotation of an *industrial pollutant* (Ge. *Schadstoff*) in texts relating to environments or environmental hazards.

Elements of Group VIII of the Periodic Table, those of *valency zero*, namely *neon*, *argon*, *krypton*, etc., constitute a third class alongside *metals* and *non-metals*. These are known as *inert elements* or sometimes by the earlier expression *noble gases* (Ge. *Edelelement*, *Edelgas*) because of their reluctance to participate in any form of chemical reaction. Molecules of inert elements consist of single atoms and remain (in a terrestrial environment) as gases, unchanged throughout time, never forming chemical compounds with any other element. These gases occur in minute concentrations in the atmosphere, though mainly in the upper regions towards the troposphere; *argon* may appear in volcanic ash. Both *neon* and *argon* have widespread applications in *lighting systems* involving discharges of electricity in gases (*neon lights*, etc.), as well as in *vacuum tubes* employed in televisions, monitors, and certain laboratory equipment. *Krypton* is expensive to produce and has no large-scale industrial applications.

11.3 Lexical Gap, Multiple Meaning

Though the German expression *edel* has an opposite *unedel*, there is no diametrically opposite concept, and English has a lexical gap. Chemists do not use expressions like *"non-inert element", *"non-noble gas" to denote *elements not belonging to Group VIII* unless, in rare cases, they themselves define them as such. Translators beware. Moreover, the expressions *edel/unedel* can have other connotations.

The morpheme *edel* is applied to metals but with a very different significance, as *metals* belong to a different area of the Periodic Table to the *inert elements*. It refers to their *oxidation tendency*. For a well-known group of metals which do not rust, the expressions *precious metal* (gold, platinum) and *semi-precious metal* (silver, mercury, copper) are acceptable translations of *Edelmetall*, *halbedles Metall*.

Edel can also apply to the reactivity of metals with acids: metals which do not react vigorously are called *non-reactive* or *noble metals* (Ge. *Edelmetall*). For the opposite extreme *unedles Metall* (lead, iron, zinc, magnesium) the expression *base metal* is used. The same term contrasts with *precious metal*.

Confusion between the multiple implications of the morpheme *edel* (inert, noble, precious, non-reactive), especially when the connotations are not clear or there is an area of overlap between the meanings *precious*, *non-reactive*, is one difficulty translators have to face. Other translation problems can occur when the adjectives *edel*, *halbedel*, *unedel* appear without a respective noun. Simple expressions with *base*, *noble*, *precious* produce violations of normal technical English, *"more base", *"semi-noble", *"half precious". Such problems must be resolved by intelligent paraphrasing:

unedel	1. highly reactive 2. tending to oxidise
halbedel	1. moderately reactive 2. with a moderate tendency towards oxidation

11.4 Solid, Liquid, Gas, Vapour

The three states of matter *gas*, *liquid*, *solid* are differentiated according to whether the molecules of the substance are able to move in three dimensions, two dimensions or not at all. The terms are used by chemists quite strictly, and lead to certain surprises. For instance *glass* is not a solid substance but a highly viscous liquid. As far as *elements* are concerned, most *metals* are solid at normal room temperature, whereas *non-metals* may exist in any of the three states. Some solids have a tendency to form *crystals*, particularly *carbon* (diamond), *silicon* and *germanium*. A few substances can pass from a solid to a gaseous state without the intermediate transition to liquid, for example *carbon dioxide* (so-called *dry ice*). This process is called *sublimation* (Ge. *Sublimation*) and is employed in refrigerators. *Iodine* and *ammonium chloride* also sublime on heating.

The term *fluid* is used by certain chemists to cover both *liquids* and *gases*. There is no direct German equivalent, and the translation *"Flüssigkeit" is likely to be very misleading. Furthermore, there is a distinction in English between *vapour* and *gas*. Substances normally in the liquid state which are induced to form suspensions of tiny particles in air or other gas mixtures are termed *vapours*. The air we breathe is a mixture of *gases* (oxygen, nitrogen, carbon dioxide) together with *water vapour*. German has expressions for the general term *vapour* (e.g. *Dampf*, *Dunst*) but seems to lack a true equivalent for the technical concept as in *sulphur dioxide vapour*, *chlorine vapour*, *ammonia vapour*. It tends to employ the expression *Gas*. Thus this is not a case of a lexical gap, simply that the terms *gas*, *vapour* do not coincide exactly with the German expressions *Gas*, *Dampf*, *Dunst*.

11.5 Solute, Solvent, Solution

In chemical processes where a solid is dissolved in liquid, the terms *solute, solvent, solution* are employed. An example is common salt, sodium chloride (*solute*) dissolved in water (*solvent*) which results in salt water or brine (*solution*). Pepper does not dissolve in water but forms a fine *suspension.* Similarly *smoke fumes* are a suspension of solid particles in gas (air) whereas *mist* is a suspension of liquid in gas. Mixtures of liquids (e.g. *alcohol* in *water*) are also *solutions.*

Again German has no 100%-direct equivalents for the terms *solute, solvent, solution.* In many cases the translations *gelöste Substanz, Lösungsmittel, Lösung* are appropriate, but English-speaking chemists tend to use the expression *solute* for a *dissolvable* substance, as well as one which has actually been *dissolved,* and German chemists normally restrict the meaning of *Lösung* to solutions involving *just* solids and liquids.

11.6 Bond, Radical

It is evident from the earlier list of chemical compounds that certain groups of atoms OH, SO_4, CO_3 appear in different substances. These combinations are called *radicals* (hydroxide, sulphate, carbonate, etc.) and reflect the structure of the compound. For substances with simple structures this has conceptual advantages, even though the chemical formulas themselves appear complex. Thus *ethylene glycol* (the main component of *antifreeze* in automobile radiators) now has the formula $HO\text{-}CH_2\text{-}CH_2\text{-}OH$ because the simple "$C_2H_6O_2$" does not indicate the symmetrical links between the radicals CH_2 and OH. There are also formulas of the type $C_6H_5CH{=}CH_2$ (*styrene*: as in *polystyrene*) where the symbol "=" indicates a double bond between the radicals concerned. But this denotation has its limitations, especially where the atoms form a *ring* — as in benzene (C_6H_6), the *spirit* used for removing clothing stains.

Certain *chemical reactions* can be generalised: *metals* dissolve in *acids* and liberate *hydrogen*; *acids* mixed with *alkalis* lead to *salt solutions.* In such cases, *radicals* move unchanged from one molecule to another.

11.7 Oxidation, Reduction

When copper is heated in air it turns black. It combines with oxygen to form copper oxide. The *oxidising agent* is oxygen itself. Copper is said to be the *reducing agent*. A similar reaction occurs with copper and chlorine, where chlorine is the oxidising agent. *Oxidation* has nothing to do with oxygen in this sense, but concerns the loss or acquisition of electrons. In both cases copper is responsible for *reduction*: copper atoms lose an electron to the oxidising agent and become negative ions, *cations*. The oxidising agent gains an electron and becomes positively ionised, resulting in *anions*. The terms *oxidation* and *reduction* also apply to reactions involving metals and acids, such as iron and hydrochloric acid, HCl. In such reactions the reducing agent is generally the *metal*, the oxidising agent (chlorine in HCl) a *non-metal*.

It often bothers translators that the terms *oxidation*, *oxidising agent* (Ge. *Oxidation*, *Oxidationsmittel*) occur in contexts involving not oxygen itself but substances whose chemical behaviour reflects one small aspect of the *behaviour* of oxygen. Chemists in both English- and German-speaking industrial environments seem unperturbed by this unusual extension of technical meaning.

11.8 Reagent, Catalyst, Inhibitor

Those substances inducing a chemical reaction itself are termed *reagents*, whereas those present simply to speed up the process and which do not themselves undergo any change are termed *catalysts*. For example, the chemical process for the manufacture of ammonia employs iron as a catalyst. Substances which slow down a chemical reaction are termed *inhibitors*. One of their main applications lies in the *rust protection* (Ge. *Korrosionsschutz*) of metallic surfaces.

The distinction between *reagent* and *catalyst* is minimal. German chemists generally use the term *Katalysator* for both. They seem to manage, by and large, without the general concept *inhibitor* too, but where it does occur the usual translation is *Passivator*. Other possibilities are *Antikatalysator*, *Inhibitor*, *Hemmstoff* according to how drastic the inhibition process is intended to be. There are catalysts which take part in *metabolic processes*, so-called *bio-catalysts*, a group which includes *enzymes*, *vitamins* and *hormones*.

Most reactions are *exothermic*. They generate heat. Even a rusty nail, which results from *slow combustion* in moist air, loses heat. In the normal engineering sense, *combustion* implies what a chemist calls *fast combustion*, for example the oxidation of phosphorus on a burning match. *Combustion* in the broad sense is defined as *exothermic oxidation*. There are also *endothermic reactions*, such as nitrogen with oxygen in the formation of nitrogen monoxide (NO). These absorb heat.

11.9 Organic/Inorganic Chemistry

Organic chemistry concerns *carbon compounds* (the substance of *living matter*) which occur abundantly in nature and technology alike. One broad class of organic compounds is the *hydrocarbons*, whose molecules consist of just *hydrogen* and *carbon*. They mainly result from industrial processes, especially in the *petrochemical industry*, which provides chemicals from fuels such as *petroleum* or *natural gas*. The hydrocarbons constitute a *homologous series* of substances whose chemical formulas conform to an algebraic sequence. The most frequent examples are the *alkanes* (methane, ethane, etc.) with the general chemical formula C_nH_{2n+2}, the alkenes (ethene, propene, ...) C_nH_{2n} and the alkynes (ethyne, propyne, ...) C_nH_{2n-2}.

Figure 10C lists a variety of *organic compounds* (see next page).

One group of organic compounds, the *polymers*, centres on industrial materials produced by linking the atoms of hydrocarbons into long *chains*, *rings* and other molecular configurations. *Polymerisation* is a major industrial branch of organic chemistry. Just as large *silicon monocrystals* (Chapter 5) are *grown* for the electronics industry, so it is possible to produce giant molecules, so-called *macromolecules*, from the alkanes, alkenes and other hydrocarbons. An example is *polythene* (originally *polyethene*). Such materials are termed *polymers*, where a *mer* or rather *monomer* corresponds to the smallest constituent, e.g. *ethene* C_2H_4. These materials form the basis of what is commonly known as the *plastics* industry. *Plastics* and *synthetic rubbers* account for half the world production of organic chemicals. *Plastics* are broadly divided into *thermoplastics* and *thermosetting plastics*. Thermoplastics soften on gentle heating but harden again; their uses include pipes, bottles and bowls. Thermosetting plastics become progressively harder on heating; an example is *bakelite* which is used in *light switches* and other *electrical fittings*.

Organic compounds are classed according to their molecular arrangement: *aliphatic compounds* contain carbon atoms linked in chains, whereas *aromatic compounds* (so-called because of their characteristic aromas) contain a ring of carbon atoms. *Benzene* (C_6H_6), the white spirit used for cleaning purposes or

Figure 10C. *Extract*

Group	Chemical Term	General Term	German
Alkanes			
CH_4	methane	methane	Methan
C_2H_6	ethane		Äthan, Ethan
C_3H_8	propane	propane	Propan
C_4H_{10}	butane	butane	Butan
C_5H_{12}	pentane		Pentan
Alkenes			
C_2H_4	ethene	ethylene	Äthen, Ethen
C_3H_6	propene		Propen
C_4H_8	butene		Buten
Alkynes			
C_2H_2	ethyne	acetylene	Äthin, Ethin
C_3H_4	propyne		Propin
C_4H_6	butyne		Butin
Alcohols			
C_2H_5OH	ethanol	ethyl alcohol	Äthanol, Ethanol
C_3H_7OH	propanol		Propanol
C_6H_5OH	phenol	carbolic acid	Phenol
Acids			
C_6H_5COOH	benzoic acid		Benzolsäure
CH_3COOH	ethanoic acid	vinegar	Äthansäure
Other Hydrocarbon Compounds			
C_6H_6	benzene	spirit	Benzol, Benzen
$C_6H_5NO_2$	nitrobenzene		Nitrobenzol
$C_6H_5NH_2$	phenylamine	aniline	Anilin
CH_2CHCl	chloroethane	vinyl chloride	Vinylchlorid
$CH_3COOC_2H_5$	ethyl ethanoate	ester	Ester
Monomer	*Formula*	*Polymer*	*Polymer, German*
ethene	C_2H_4	polythene	Polyethen
styrene	$C_6H_5CH{=}CH_2$	polystyrene	Polystyrol
chloroethane	CH_2CHCl	PVC	PVC
tetrafluoroethene	C_2F_4	teflon	Teflon

removing stains, is an *aromatic hydrocarbon*, whereas the *paraffins*, the group now more correctly known as the *alkanes*, are *aliphatic*. Terms such as *chlorinated* or *fluorinated hydrocarbons* describe materials where one hydrogen atom has been replaced by a *chlorine* or *fluorine* atom. Such materials (e.g. PVC, Teflon) are difficult to dispose of and are seriously detrimental to the environment. A term denoting the general combined group *chloro-fluoro-hydrocarbons* recently entered general language too: *CFC's* (chlorofluorocarbons, Ge. *Fluorchlorkohlenwasserstoffe, FCKWs*).

The main branch of *inorganic chemistry*, which is currently attractive to industry, centres on compounds involving *silicon* (Ge. *Silicium*), the so-called *silicones* (Ge. *Silicon*). These are used in lubricants, raincoats, shock absorbers, and a wide variety of unrelated applications. Silicones are composed of molecular chains of alternate *silicon* and *oxygen* atoms. Other inorganic compounds produced on a large scale include *ammonia* (NH_3), the various acids (HCl, H_2SO_4, H_3PO_4, etc.) used in the fertiliser industry, *hydrogen peroxide* (H_2O_2) used in washing powders and bleaching agents, and *calcium hydroxide* $Ca(OH)_2$, which is employed in disinfectants as well as for neutralising waste acids in industrial effluent.

11.10 Recycling, Reprocessing

Whereas governments impose strict controls on nuclear materials, they are notoriously lax when it comes to supervising the disposal of *chemical wastes.* Just as chronic air pollution and murderous smogs led to higher and higher factory chimneys, so the pollution of soil, ground water, rivers and estuaries now leads to *chemical dumping* farther and farther out to sea. The gradual destruction of the complete biosphere of our planet by chemical pollutants means that much of current technical literature is concerned with minimising the dangers, particularly where the substances concerned are banned in many countries. The alternative is to store the chemicals in a safe place for *reprocessing* at some time in the future when industry has acquired the necessary expertise. Thus, terminological distinctions introduced in Chapter 4 in connection with *nuclear waste* apply in this field too: *landfill, disposal site, dump; reprocessing plant, storage site, repository; residual, transitional* and *terminal waste.*

Many *disposal sites* (Ge. *Deponie*) from former periods urgently need to be cleaned up, especially in Eastern Germany and various Eastern European countries. Thus methods of *soil decontamination* (Ge. *Bodensanierung*) are currently in great demand. Similar techniques are employed for separating *mixtures* of dangerous chemicals at chemical reprocessing plants. One ingenious method of collecting and removing *heavy metals*, such as *cadmium*, from the soil is to sink gigantic electrodes into the contaminated area and apply an electric field. By virtue of their positive charge, the heavy metal ions eventually migrate towards the cathode, where they are collected and disposed of. Ions of *cyanide* and other materials which bear a negative charge migrate to the anode. Other methods of separation involve the use of *chemical bonding agents* which prevent *noxious* or *toxic substances* from entering the biocycle and enable them to be washed out of the soil as a separate *sludge*. Microbiology has applications here too, as special bacteria or microbes can be bred which feed on *hydrocarbon* compounds such as *machine oil*.

Some chemicals can only be broken down at special *incineration plants* where the waste is ignited at temperatures of around 1500C. Filters are required to deal with the *flue gases* (Ge. *Rauchgas*) and to remove heavy metals, acidic and other harmful substances. Many different plastics resulting from household waste collection are also disposed of in this manner. In view of the enormous heat generated by the *incineration furnaces*, attempts are made to combine them with *steel works* and other industries requiring vast heat sources or *blast furnace* facilities.

Unit 12

Electronics

By the middle of the twentieth century, pioneering research in the field of Materials Science was beginning to pay dividends. Revolutions in technology were taking place in the textile industry due to the work of chemists sponsored by large industrial manufacturing concerns, with terms like *rayon* and *nylon* becoming household words. The plastics industry was discovering its great potential, and by the sixties new *metal alloys* with unprecedented material properties were in great demand by the *civil aviation* and *aerospace* industries. Amidst this hive of research activity, one important technique discovered by materials scientists and one tiny innovation from 1948, neither of which received much acclaim at the time, were to have ramifications later affecting the whole of technology itself: the growing of *semiconductor crystals* (Ge. *Halbleiterkristall*) and the manufacture of the first *transistor*.

This unit provides a broad summary of two very detailed chapters of the book: Chapters 5 and 6. There is no room for lengthy *contents* lists, or details of the relevant glossaries and thesauri. This time the reader is plunged straight in.

12.1 Early Electronics

One general problem for translators in this area is that linguists react rather slowly to the technological changes, and published dictionaries are inclined to retain unnecessary terminology from earlier periods. There are still a few electronics dictionaries (and data banks) which contain entries like:

Kondensator	capacitor, condenser
Plattenspieler	gramophone, record player
Radio	radio (Am.), wireless (Br.)
Röhre	tube (Am.), valve (Br.)

with no indication that some target language expressions refer to objects that are completely obsolete. This section examines terminology which has altered or modified its significance during the gradual evolution of the electronics industry. It spans the period from the time of the early circuit technology centred on *thermionic tube devices* (Ge. *Röhre*) to the equivalent technology involving *semiconductor devices* which has led to the *microelectronics* industry.

The nineteen-twenties to the fifties witnessed the first upsurges in the electronics sector, especially in regard to the large-scale marketing of expensive household gadgets known as *wirelesses* and *gramophones*, and eventually also *televisions*. The sales push continued in the sixties, but the new electronic components employed differed radically from those of the early years, due mainly to the advent of cheaper, smaller, more flexible components produced by the semiconductor industry, among which were *transistors*. Wirelesses evolved into so-called *transistor radios* (known for a brief period in the sixties by the misleading colloquial designation of "transistors"); bulky gramophones evolved into *record players*.

Not only did the terminology of the finished market product change (*wireless* to *radio*, *gramophone* to *record player*, etc.). So did the terminology of the individual *components*, at least in the English-speaking areas of the world. Clumsy metallic devices known as *resistances* and *rheostats* were replaced by small elegant components known as *resistors* and *potentiometers*. Costly high-precision mechanical arrangements of *parallel plates* constituting the *condensers* of early wirelesses were replaced by miniature versions operating along different lines and known as *capacitors*. Germany experienced these same changes in technology but retained its older terminology (i.e. *Widerstand*, *Kondensator*, etc.), merely shifting the focus of meaning to the new devices.

The term *diode* underwent a complete semantic shift in both languages, from the expensive, dangerous, high-voltage *thermionic devices* of the early gramophone, so-called *diode valves* (Ge. *Diodenröhre*) to the tiny, silent, harmless *semiconductor* equivalent (Ge. *Halbleiterdiode*). But *triode* has retained its original meaning (Ge. *Triodenröhre*). This is partly because its nearest semiconductor equivalent, the *transistor* operates in a very different manner, and partly because unlike thermionic diodes, triodes have certain features and properties which are still useful to the *live-music* industry and which cannot be reproduced by semiconductor devices. Thus triodes are still manufactured. But they are no longer referred to as *valves*; this former British expression has largely been replaced by the competing American designation *tube*.

Circuit-design technology based on Tube Electronics has reached an evolutionary dead-end. That of Semiconductor Electronics is continually breaking new ground, as well as new records, in regard to *micro-miniaturisation.* For most of the e-book (i.e. unless otherwise stated), the expression *Electronics* implies *Semiconductor Electronics.*

12.2 Semiconductors, ICs

By the late nineteen-sixties the devices most frequently employed in circuits were: *transistors, diodes, resistors* and *capacitors.* Extensive use was made of these components, as opposed to *inductors, thermionic tubes,* and other devices which had a longer history of development, because the former were much smaller, more reliable and very much cheaper. Many standard circuits were redesigned during this period to avoid the more expensive traditional components completely, in some cases regardless of the additional complexities involved. Semiconductor materials were used to make transistors and diodes but it soon became possible to manufacture resistors and capacitors from the same materials as well. It was then just a small step to the manufacture of complete *integrated circuits* (*IC's*) on single minute pieces of semiconductor material (*chips*); continued refinements in technology led to *miniaturisation* and *micro-miniaturisation,* and subsequently to the *micro-chips* of everyday life which control our computers, the *fuel consumption* of our motor vehicles and, in later years, the *pacemakers* (Ge. *Herzschrittmacher*) controlling our heartbeats.

Semiconductor materials have an electrical conductivity lying between that of *conductors* and that of *insulators.* This is the suitable introductory definition given in many engineering textbooks. In practice, however, the conductivity of semiconductors is far more dependent on electrical and thermal environments and on *impurity concentrations* than is the case for most other materials, and this is what really distinguishes them from conductors and insulators. *Impurity atoms* of specific elements are deliberately injected into pure *monocrystals* of silicon or germanium in order to achieve particular conductivities, a process known as *doping* (Ge. *dotieren*). Pure materials are termed *intrinsic semiconductors* and those which are doped *extrinsic.* A sample of extrinsic semiconductor material which has a surplus of *negative* charge carriers due to doping is referred to as *n-type material* (Ge. *n-Halbleiter*); extrinsic semiconductors with a surplus of *positive* carriers are designated *p-type.* At this point, readers with no previous translation experience in the field of Electronics may find it helpful to first browse through certain *disk illustrations*, those in the (thumbnail) sections Basic Electrical Engineering and Materials Science/Semiconductors, with special attention to the illustrations: *Crystal Lattice, Semiconductor Material.*

A region of a semiconductor where there is an abrupt transition from p-type to n-type material is referred to as a *junction region* and the interface itself as a *pn-junction.* The junction region is just a few *microns* (thousandths of a millimetre) wide but consists of a smooth gradual *transition* from material with a high acceptor concentration to that with high donor concentration. It is a surprise to some people that the terms *transition* and *junction* are contextually synonymous in this field: the *abrupt junction* in the physical sense constitutes at the same time a *smooth electrical transition.* German employs the expression *pn-Übergang* for both shades of meaning.

It could be said that the entire electronics industry really originates from the unique properties of the pn-junction, which provides a smooth silent mechanism for *current control* and can be used to block current completely. Semiconductor devices containing a single junction (two layers of material) are known as *diodes*, and can be used to ensure that current flows in one direction only, so-called *rectification* (Ge. *Gleichrichtung*). Devices with two junctions constitute *transistors*, and are used for switching applications or amplification purposes (Ge. *Verstärkung*). A more elaborate type of electronic switch, the *thyristor*, has three junctions.

12.3 Conduction, Bonding

The conductivity of a piece of semiconductor material depends a lot on the degree of *purity* of the crystal, the *perfection.* Once a *pure crystal* is obtained the conductivity can be closely controlled by one or more of the following means: *application of heat, incidence of light, impurity injection.* Most semiconductor devices (diodes, transistors, thyristors, IC's, etc.) depend entirely on impurity injection and are sold in *light-proof* metallic cases designed to dissipate any internal heat produced. Exceptions to this are the class of devices used as *sensors* in fire alarms, oven lamps, burglar alarms, etc. *Thermistors* (thermal resistors) are the main heat-sensitive semiconductor device, and *LDR's* (light-dependent resistors) the main photo-sensitive component. Pressure-sensitive (piezo-electric) devices also exist and there are semiconductors which detect radioactivity.

The conductivity of pure silicon monocrystals, so-called *intrinsic conductivity*, arises mainly from the liberation of electrons by *thermal agitation* of the lattice atoms. It is thus very dependent on temperature. *Extrinsic conduction* (Ge. *Störstellenleitung*) differs from its counterpart *intrinsic conduction* (Ge. *Eigenleitung*) in that the liberation of a fixed number of charge carriers is effectively guaranteed by the impurity concentration (the *donor/acceptor* concentration), though other factors are involved which depend on the material itself. *Extrinsic conductivity* is relatively independent of ambient temperature.

Expressions like *Atombindung, Bindungselektron, Bindungskräfte* occuring in this field are translated into English by *atomic bonding, bonding electron, bonding forces*, rather than expressions involving a similar but unrelated concept from Nuclear Physics: *binding*. The distinction appears in Chapters 3 and 4 too. *Bonding* implies the physical process by means of which materials are held together by virtue of *electrons* sharing the *valence shells* of their nearest neighbours. *Binding* concerns the forces of cohesion present among *protons* or *neutrons*, which prevent *nuclei* from disrupting. There are other *binding forces* which constrain *electrons* to remain within particular orbits around their nuclei. Both *bond* (participle: *bonded*) and *bind* (participle: *bound*) correspond to the German *binden (gebunden)*. Translators beware.

12.4 Impurity, Contaminant, Pollutant

In chemical engineering the term *impurity* (Ge. *Fremdstoff*) implies a substance which does not belong to the material involved and which may *adversely* affect its properties. When the material is an alloy, polymer or liquid substance, a stronger alternative expression is *contaminant*. If the context concerns air, water or some other essential aspect of human, animal or plant life affected by a contaminant the contaminating substance is usually called a *pollutant* (Ge. *Schadstoff*). In the context of semiconductor engineering, however, the technical expression *impurity* has none of these negative connotations.

US-based semiconductor scientists, such as Van Vlack, Azaroff, Brophy, who may have had direct access to German scientific literature, introduced the term *foreign substance* in the sixties and seventies as an alternative to *impurity*. But the expression seems to have died a natural death. Thus, terms like *Fremdstoff, Fremdatom, Dotierungsgrad* are rendered in English as *impurity, impurity atom, impurity concentration*, whereas concoctions involving the expressions *foreign, contaminating, polluting* are generally avoided unless they imply the broader significance of an *undesirable* chemical impurity.

12.5 Resistance, Capacitance

Faced with the German term *Kondensator* in a text describing the internal components or configuration of devices present in an integrated circuit, it would be quite ridiculous for the translator to substitute the ancient tube-electronics expression *condenser.* The object concerned has the function of a *capacitor.* Yet an expert on *semiconductor engineering* who specialises in *IC manufacture* might refer to the minute area of silicon in question not as a *capacitor* but as a *capacitance.* The engineer concerned regards the IC as consisting of a multiplicity of differently doped *regions* or *domains,* rather than discrete components; *capacitative domains* are dubbed *capacitances.* Likewise, *resistive domains* are not necessarily called *resistors*; the terminology has gone full cycle: *resistances.*

This does not mean to say that the substitutions *capacitor, resistor* are wrong in the above case. They depend on the customer's individual preference. Understanding is not impaired, as it would be by totally false substitutions, such as *capacity, *condenser, *impedance. These examples are mentioned merely as an indication of how technical language, like natural language, adapts itself to the situation at hand.

12.6 Reactance, Impedance

Not all *electronics* involves ICs. For many consumer applications, individual resistors, capacitors, transistors, etc. have to be employed, so-called *discrete components* — as distinct from *modules,* of which the IC is merely one variety. *Circuit components* are termed *passive* if there is a *linear* relationship between the current conducted by a particular device and the voltage applied across it. The three main passive devices are: *resistor, capacitor, inductor.* Other components which do not behave in this manner are termed *active. Tubes, diodes, transistors, thyristors* are examples of active components.

In passive components used under ac conditions the mean ratio of *voltage* to *current,* the so-called *impedance,* is constant, that is to say it does not change according to the voltage itself or to the current applied. The *impedance* of a *resistor* is independent of frequency and equals the *resistance* of the device. For other devices, the value varies according to the *frequency* of the *ac signal,* but impedances of passive devices can be determined at a given frequency by a relatively simple calculation.

For inductors and capacitors the parameter *resistance* is irrelevant except to specify minimal side effects, such as *coil resistance* (inductor), *dielectric resistance* (capacitor), which result in energy losses due to heat. *Reactance* can be calculated from the following simple formulas: *inductive reactance* = ωL; *capacitative reactance* = $1/\omega C$. The symbols *L* and *C* correspond to the values of *inductance* and *capacitance*; ω (omega) represents the *angular frequency* of the signal, a quantity measured not in *Hz* but in the unit $radian.sec^{-1}$.

Circuit designers can compensate for undesirable inductive effects by inserting *capacitors*, and for capacitative effects by employing *inductors*. Hence, despite their very different appearance and function, there is a term covering both devices: *reactances* (Ge. *Blindwiderstände*). The *impedance* of a *reactive component* (a capacitor or inductor) is obtained by a sophisticated calculation involving *vector addition* of *reactance* with *dielectric* or *coil resistance*. The parameter is specified in *magnitude* and *phase*.

12.7 Transducer

There is a small group of semiconductor devices whose *resistance* varies linearly according to one feature of the *external environment*. These special *semiconductor resistors* may be *heat-sensitive* (thermistor), *light-sensitive* (LDR) or *pressure-sensitive* (piezo-electric resistor). The devices have important industrial as well as domestic applications: *burglar alarms*, *fire alarms*, *oven warning lamps* or *pick-ups* (Ge. *Tonabnehmer*) for acoustic musical instruments; they convert different forms of energy into electrical energy and are referred to collectively as *transducers*.

There are transducers for sensing optical, acoustic, thermal, magnetic and mechanical energy (light, sound, heat, magnetism and pressure) and semiconductor transducers exist which respond to radioactivity. Some devices operate in the reverse direction, in that they convert *electrical signals* into *optical* ones. The most famous is the *LED* (light-emitting diode) found in the *digital displays* of calculators, digital watches, digital meters. Thus some transducers are *sensors*. Others are not.

The basic meaning of *transducer* in engineering is a device that converts energy from one form into another. The implication attached to the term in the field of Electronics is a device that converts *non-electrical signals*, such as pressure changes, fluctuations in light intensity, temperature variations, into *electrical ones*, and vice versa. Thus the normal meaning of *transducer* covers the set of devices known as *control devices* (Ge. *Steuerelement*), for instance *light sensors, pressure sensors, thermistors* and *magnistors* (Figure 6A), as well as the counterparts *light-emitting diode/transistor (LED/LET)*. Theoretically, the definition should also cover light bulbs and loudspeakers but in practice it usually does not.

German seems to lack a true equivalent for *transducer*. It has the term *Meßwertumformer* as a roughly approximate translation, but this is too clumsy and the concept is usually rendered as *Umformer, Umwandler* or *Wandler*, despite the fact that these terms have other meanings in Electrical and Electronic Engineering: *Spannungsumformer* (dc/ac voltage converter), *Bildwandler* (image converter), *Phasenumformer* (phase inverter). There is a term *Transduktor* but its usage has different connotations and is mainly restricted to the field of Electromagnetism.

12.8 Bias, Operation, Mode, State

Normally, in technical English, *devices* are *"operated"*, whether electrical or not, but there is a second electronics term with a more specific meaning, namely *bias*, which has implications on the *state* of the device concerned and the *mode* of operation. Ordinary dictionaries tend to suggest the German equivalent *Betrieb* for the first two conceptions and *Zustand* for the other two, which is of little use to translators confused by these terms. The examples below may provide assistance.

A *diode*, a device that conducts current in one direction only, must be *biased* in order to provide the desired effect, in other words there has to be an appropriate difference in *potential* between the two electrodes. It can be *operated under forward bias* (Ge. *Polung in Flußrichtung*), in which case the diode is in the *conducting mode*, or *under reverse bias* (Ge. *Sperrichtung*), as in the case of zener diodes, which are operated in the *non-conducting mode.*

Similarly, a switching transistor can be *biased* so that it is in the state of being *ON* (Ge. *EIN-Zustand*). A sudden reduction or cessation of the *base current* switches the transistor into the *OFF-state.* Thus, the same transistor is operated in the *ON-* or the *OFF-mode* at different times. There is also an intermediate stage (neither OFF nor ON) known as the *amplification mode.* Whether expressions like *"im Durchlaßbereich"*, *"in Flußrichtung"* are to be rendered in English as "in the forward mode", "under forward bias" or "in the ON-state" really depends entirely on the context concerned and on the device itself. A third device, the *thyristor*, employs a mixture of this terminology.

Translators come up against a brick wall when faced with such problems. In the current absence of large-scale, systematically organised collocation dictionaries, the only solution is to study appropriate technical literature first-hand in both languages before attempting translations.

Many engineering fields have developed from branches of Experimental Physics which may have existed for decades or even centuries before practical or technological applications emerged. Newton's clear-sighted conception of Mechanics no doubt seemed as abstract and esoteric to seventeenth century scientists as Einstein's theories of General and Special Relativity appear to many people today. But not even Newton himself could not have envisaged the many areas of Mechanical Engineering which were to arise two hundred years later, encompassing *steam locomotives, petrol engines* and the construction of the Eiffel Tower, all of which depended on practical applications of his three basic laws and extensions of his pioneering mathematical skills in what is now known as *Newtonian Calculus* (Ge. *Integral/Differentialrechnung*). Similarly, Faraday and other pioneers in the field of Electricity could not have foreseen *CD-players, mobile phones, personal computers, internet libraries* or *electronic synthesizers.*

The field of *Electronics* has a short history, but one of rapid development, and it is definitely here to stay. Despite the complexities, translators must learn to cope with it. The detailed sections, subsections, glossaries and thesauri of Chapters 5–7 coax the inspired reader gradually through the initial stages.

UNIT 13

Technical Grammar

The early chapters employ technical expressions like *capacities*, *impedances*, *resistances*, which seem to abide by different grammatical rules to their natural language counterparts: *capacity*, *impedance*, *resistance*. Just as NCNs like *coffee*, *beer*, *cheese* adopt properties of CNs when they imply "*cup of coffee*", "*bottle of beer*", "*type of cheese*" so the meanings of technical terms change slightly when their grammatical categories are altered. In such cases, technical CNs often correspond semantically to measurable parameters or mathematical values associated directly with scientific concepts, the latter being NCNs. Native-speakers with literary rather than technical backgrounds tend to frown upon this apparent misuse of language, but it is perfectly natural and often, for translators, unavoidable. This unit takes a closer look at grammatical differences between technical and natural language.

13.1 Shades of Meaning

Consider the following statements, which could appear in fields such as Physics, Electronics, Nucleonics or Chemical Engineering:

1. The total *kinetic energy* of the particles, i.e. the sum of their individual *kinetic energies*, amounts to *150 MeV*.
2. The total *mass* of the particles, in other words the sum of the *masses* of the individual nucleons, amounts to *238.029 a.m.u.*
3. The *charge* of the nucleus is balanced by the *charges* of the rotating electrons.

All are well-formed, perfectly natural statements in an engineering context. The first could refer to *ions* in a *vacuum tube*, the second to the *uranium atom*, the third to any *non-ionised atom*. They illustrate the same semantic contrast. In this respect, technical language differs from general language, where statements like:

*"The *weight* of the building is determined by the *weights* of the individual bricks."

are considered *non-grammatical.*

It is difficult to say whether the meanings of *kinetic energy, mass, charge* are different in the contexts concerned, or whether the contexts themselves impose slight extensions upon the basic meanings. But there are cases where meaning in context changes very radically. The following example illustrates this aspect of specialised language and its implications on technical translation quite vividly:

1. The wire itself has *resistance.*
2. The wire itself has *some resistance.*
3. The wire has *a resistance* of 0.05 ohms.
4. The wire is equivalent to *a resistance* connected in series.
5. The *resistances* of the increments of dielectric material can result in a fairly substantial *shunt resistance.*

Statement 1 (NCN) implies that the resistance of the wire cannot be neglected, and is tantamount to saying *the wire will get (slightly) hot* or *some energy will be lost in the wire itself.* Statement 2 (NCN) implies that the wire resistance is *small* relative to other resistances in the given context. Statement 3 (CN) implies the parameter (physical quantity) *resistance,* whereas statement 4 (CN) refers to the entity *resistance* itself, for which (with a slight shift in semantic emphasis) the expression *resistor* could be substituted. In statement 5, the term *shunt resistance* signifies an unspecified measurable parameter (as in 3) and *resistances* denotes the entity itself (as in 4), though here the substitution *resistor* is impossible.

In the same way that familiarity with *general language* and attention to the context enables the listener to distinguish the utterance *some cheese* from *a cheese* and to determine whether the latter means *a type of cheese, a piece of cheese from a selection on a plate* or *one of a set of identical pre-wrapped cheese samples in a supermarket package,* so familiarity with the subject matter enables engineers and hopefully also translators to distinguish different *extensions* of basic engineering terms like *resistance, charge, energy, mass, force, power, tension,* etc.

13.2 Entity, Property, Parameter

The discussion above reveals five marginally different interpretations of the technical term *resistance,* some being CNs others NCNs, but all of which correspond to the German *Widerstand*. As *Widerstand* itself is a polyseme and is translated by *impedance, reactance, resistivity*, etc. (Chapter 2), the result of incorporating separate entries for separate sub-interpretations in the TPD could look rather ugly and remain unrevealing. Furthermore, *charge, current, voltage* and indeed the overwhelming majority of physical quantities discussed in the opening chapters have secondary interpretations, which could make the Thesaurus unnecessarily bulky. The dictionaries therefore employ certain simplifications in this respect, and distinguish just three types of entry: CN, NCN, CN/NCN.

The various implications of *resistance* in the examples mentioned are partly due to the influence of articles or quantifiers: *some, any, a/an*, etc. The number of different concepts is reducible to three: an *entity* (with implications comparable to those of *resistor*), a *property* characterising the entity, and a *parameter* which characterises the property and is specified in terms of *units* (ohms). The *parameter* is linked to a *value* such as "50 ohms". Parallel semantic distinctions occur for other electrical quantities too: *inductance, reactance, impedance, charge*. In such cases: the *entity* is a CN, the *property* a NCN, the *parameter* also a CN.

Some terms denoting physical quantities apply only to property and parameter, e.g. *energy, power, voltage*. These are registered as parameters, i.e. CNs, in the Thesaurus, whereas the TPD indicates their dual function by the label CN/NCN. Other engineering terms denote parameters only, for instance *resistivity* is a parameter relating to the property *resistance*. These are CNs. Others denote only properties, for instance *inertia*, and are NCNs.

13.3 Hyponymy, Countability

Chapter 1 reveals that the concept *force* has many manifestations. These are expressed by compound terms in German (mostly containing *-kraft*) but by different terms in English: *thrust, traction, tension, weight*, etc. The reader might logically deduce that these too are CNs, like the superordinate technical concept *force*. Though this is true in many cases, it does not apply to all. Consider the example *gravity* (Ge. *Schwerkraft, Massenanziehungskraft, Gewichtskraft*):

1. The meteor has *gravity* of its own.
2. (*)The meteors have *gravities* of their own.
3. *The meteors have *gravity* of their own.
4. The meteors exert *gravitational forces* of their own.

Sentence 1 above might appear in a translation of a text on Astronomy. In technical English, *force* is normally a CN but *gravity* is not, even though there is a direct hyponymic relationship between the two concepts. Problems occur when the translator is faced with a *plural version* of this sentence in the source language.

A few physicists and engineering scientists working continually in this field may feel intuitively that the transition of *gravity* from NCN to CN has taken place. But most astronomers would hesitate to write *"a gravity of its own"* and shudder at the use of *"gravities"*. Sentence 2 is therefore only marginally acceptable. The third sentence is grammatically acceptable but not semantically: it gives the mistaken impression that each meteor has the same gravitational field. Sentence 4 resolves the problem by paraphrasing.

In the same manner, the concept *friction*, which has the *dimensions* of *force* but denotes the *resultant* of the frictional forces acting between two surfaces, remains a NCN: * *"frictions"* is quite wrong. Similarly, *energy* and *work* have the same units and dimensions and are semantically very closely related in all fields of science and technology. Yet the technical parameter *energy* is normally a CN, *work* (like *friction*) is always a NCN: * *"works"* is impossible in this context. A similar grammatical contrast concerning the closely related concepts *power* (CN/NCN) and *heat* (NCN) exists in the field of Electronics.

Although translation problems involving countability considerations are restricted to just a small set of lexemes, e.g. *heat, inertia, resistance, tension,* such terms occur repeatedly throughout technology. Conventional technical dictionaries, including on-line facilities, currently provide little assistance in this respect.

Unit 14

Technical Thesaurus

The Technical Thesaurus provides explanations for the scientific and engineering expressions discussed in Volume One or listed in the TPD. The set of descriptors used to indicate interrelationships among the terminology of the Thesaurus appears in the disk section *Dictionary Symbols*. For ease of reference, the abbreviations are:

a:	associated with ...
co:	consist(s) of ...
ct:	contrasted with ...
cv:	covers the concept(s) ...
d:	defined as/designates ...
ex:	(typical) example ...
m:	a (measurable) parameter characterising ...
p:	part of ...
s:	synonym for/abbreviation of ...
t:	a type of ...
tu:	typical unit ...
u:	used in connection with/used for ...

cs/ps/nps are subcategories of *s*, denoting *contextual/preferred/non-preferred synonym* respectively.

The Thesaurus has the following objectives:

i. to specify concepts associated with the basic terminology of engineering;
ii. to indicate homonyms in English technical literature;
iii. to distinguish polysemes in technical English and clarify those of technical German;
iv. to indicate hyponymous relationships among English terminology;
v. to provide access to German compounds in the TPD from English.

These objectives are demonstrated with examples in the sections below, which also illustrate use of the descriptors.

14.1 Concept Specification

Some moderately difficult terminology for inexperienced translators appears right at the start of the book. Chapter 1 discusses different *forces* encountered in engineering applications: *thrust, tension, compression, traction*, etc. Chapter 2 differentiates terms like *impedance, resistance, reactance, resistivity*, which in German are interpreted by compounds involving *Widerstand*. It also mentions *charge, current, voltage, potential, emf* (Ge. *Ladung, Strom, Spannung*). Basic concepts such as these and many expressions from other chapters reappear with appropriate definitions in the Thesaurus.

Once the reader has mastered the thesaurus technique, in the course of reading and digesting Volume 1, the way is clear for the identification of terms like *caliper, pinion, tailstock* and other concepts appearing only in the second volume.

caliper	AUTO	*Bremssattel,-m*
u: disc brake.		
pinion	MECH	*Ritzelrad,-n*
t: small gear meshing with a larger gear or rack.		
tailstock	MECH	*Reitstock,-m*
p: lathe.		

The examples reveal that the term *caliper* is used in connection with *disc brakes* in the field of Automobile Technology (AUTO), that *pinion* is a type of *small gear* and *tailstock* a part of a *lathe*, both terms occuring in Mechanical Engineering (MECH).

14.2 Homonymy

The second function of the Thesaurus is to distinguish technical homonyms in English, such as:

pitch (u: road surfacing)	CONS	*Pech,-n*
pitch (u: acoustic wave)	ACU	*Tonhöhe,-f*
resolution (u: monitor, TV)	OPT	*Bildauflösung*
resolution (u: forces)	PHYS	*Zerlegung*

Even the reduced definitions above indicate that the concept *pitch* used in connection with *road surfacing* is quite different from that associated with the acoustics term *pitch* (i.e. *voice pitch*, etc.). Similarly the optical significance of *resolution* is not the same as that of the Physics term *resolution* used in connection with *forces*.

It is easy to locate homonyms by virtue of the alphabetic arrangement but the actual distinction *homonym/polyseme* is often blurred. For example:

quantity (u: number of items)	GEN	*Menge,-f*
quantity (cs: physical quantity)	PHYS	*Größe,-f*

The above entries show that the general meaning (GEN) of *quantity* differs substantially from that of the Physics term (PHYS) used in Chapter 1. Nevertheless, the important thing for the translator is not to distinguish homonyms from polysemes, but to select the appropriate target-language equivalents. The Thesaurus provides the semantic clues needed for concept differentiation.

14.3 Polysemy

The Thesaurus differentiates polysemes of the English technical language in two different ways. The first method is similar to that for homonyms and follows from the alphabetic arrangement of the entries. For instance:

body ASTR *Himmelskörper,-m*
cs: celestial body; ex: planet, comet, asteroid, meteorite.
body AUTO *Karosserie,-f*
ct: chassis, engine; cs: bodywork.
body NAUT *Boots-/Schiffskörper,-m*
p: boat, ship, submarine; u: hull.
body PHYS *Körper,-m*
cs: object; ex: body at rest, moving body, celestial body.

The fact that the technical term *body* has various meanings is apparent, simply because the four entry terms occur adjacently and have different German equivalents. Perusal of the thesaurus helps the reader to understand the polyseme.

Sometimes a contextual synonym warrants a separate entry itself, for instance *bodywork*. The second entry extends the information of the first.

bodywork AUTO *Karosserie,-f*
ct: paintwork, woodwork, chromework, rubberwork; cs: body.

The other method by which polysemes are distinguished in the Thesaurus is used when the term itself is easily understood, but the English polyseme is barely recognisable and might otherwise be overlooked. Rather than encumber the Thesaurus with unnecessary definitions, there is a second type of Thesaurus entry — one that looks more like those of the Polyseme Dictionary — which lists "specimens" of the various interpretations. For example, the entry *clip*:

crocodile clip	ELEC	*Krokodilklemme*
spring clip	MECH	*Federspange*
paper clip	OFF	*Büroklammer*

Other examples are *gap*, *load*:

electrode gap	IGN	*Funkenstrecke*
energy gap	SEMI	*Energielücke*
contact breaker gap	IGN	*Unterbrecherabstand*
axial load	AUTO	*Achslast,-f*
cargo load	AERO	*Frachtladung,-f*
payload	ROCK	*Nutzlast,-f*
tensile load	CONS	*Zugbelastung*

These entries provide access to other entries at *Klemme*, *Abstand*, *Last*, etc. in the TPD which specify the root concept implied.

Different specialised terms, such as *solution* (u: mathematical equation) and *solution* (a: chemical laboratory), warrant separate entries in the Thesaurus, even if the polyseme happens to converge in German (i.e. *Lösung*). There are also separate entries when the English concept is roughly the same but German happens to express it differently, for instance *rotational speed* (Ge. *Drehgeschwindigkeit*) and *engine speed* (Ge. *Drehzahl*).

14.4 Hyponymy

Hyponymy is indicated in the Thesaurus in three ways:

i. *directly* — by including the hyponyms as separate entries;
ii. *indirectly* — by virtue of the relational descriptors employed;
iii. *selectively* — by providing access to the hyponymns in the TPD.

i. ***direct indication***

For compounds whose exact meanings are difficult to establish from the constituents, for instance those derived from the electrical term *circuit — switching circuit, resonance circuit, open circuit, short circuit, closed circuit*, full thesaurus definitions are provided. These may be true hyponyms (*switching circuit*, t: *circuit*) or derived meanings, only indirectly related to the root term (*short-circuit*, a: *circuitry, wiring, electronic device*). German equivalents may differ when different hyponymy is involved (*Schaltkreis, Kurzschluß*).

ii. ***indirect indication***

The second method of hyponymy indication involves the descriptor *cv* (*covers the concepts*), and is apparent in the example:

control ELEC *Bedienungselement,-n*
cv: knob, switch, button, key.

This entry provides the information that the terms *knob, switch, button, key* are hyponyms of the electrical expression *control*, in other words each is a type of control. The global term may be used collectively (i.e. *controls*) to cover the set of hyponyms, it may appear in compounds (e.g. *control knob, button control*), or under certain conditions it may occur as a substitute for one of the hyponyms.

iii. ***selective indication***

Not all hyponyms warrant separate thesaurus entries, especially those whose German equivalents appear adjacently in the TPD anyway. Rather than simply reversing and repeating similar lists to those in the Polyseme Dictionary, the Thesaurus economises. Compounds such as *alphabetic character, numerical character, volatile substance* are not included, as their meanings are fairly self-evident once the root term is established and German equivalents are locatable at the corresponding TPD entries.

14.5 Hierarchic Relations

It is useful, when using the Thesaurus, to have a piece of rough paper handy. This enables the translator to jot down unfamiliar terms in the form of conceptual hierarchies which enable the concepts themselves to be memorised more easily. This can either be done in the form of tree diagrams or, if the Thesaurus deductions are to be stored in a notebook or data file, the representations introduced in Volume 1 can be employed. The descriptors *t* (*type of*), *p* (*part of*), *co* (*consists of*) and *ex* (*examples*) are particularly amenable to hierarchic representations. This is demonstrated in the examples following:

magnetic sensor	ELNC	*magnetfeldabhängiges Element*
t: control device; ex: magnistor.		
motor	ELEC	*Motor,-m*
co: field windings, armature, brushes		
needle valve	AUTO	*Schwimmnadelventil,-n*
p: carburettor; u: float, float chamber.		

The examples reveal that *magnetic sensor* is a generic hyponym of *control device*, and that *magnistor* is a hyponym of *magnetic sensor*. Also that the concepts *field winding, armature, brush* are parts of a *motor*, and *needle valve* of a *carburettor* — information verifiable incidentally from the disk illustrations *Starter Motor, Carburettor.*

This technique for denoting relationships among terminology is less explicit but nevertheless analogous to hierarchic arrangements. The above examples lead to the configurations:

1	control device	2	motor
11g	magnetic sensor	21p	field winding(s)
111g	magnistor	22p	armature
		23p	brush(es)

3	carburettor
31p	needle valve
32a	float
33a	float chamber

where the symbols *g, p, a* denote the semantic relationships *generic, partitive, associative.* The hierarchies can be extended in any direction ("upwards", "downwards" or "sideways") simply by looking up related terms in the Thesaurus. This reveals for instance that a *carburettor* is part of the *fuel system* of a motor vehicle and that other examples of *control devices* are *light, heat* and *pressure sensors.*

14.6 Contrast

The descriptor *ct* corresponds to the semantic relation *contrast.* It is illustrated by the entry:

conventional ignition	AUTO	*Standardzündung,-f*
ct: electronic ignition.		

which provides the information that the term *conventional* in connection with automobile ignition systems contrasts with *electronic*. Subsequent reading (e.g. Chapter 8) reveals the fundamental distinction that conventional ignition systems employ *inductive discharge* (the spark energy derives from the ignition coil) whereas electronic systems employ *capacitative discharge* (the energy derives from a capacitor). This may or may not be not important to the translator at this stage. The main thing is that the contrast is made and the "key" provided for further access to other information sources.

To take a simpler example:

element CHEM *chemisches Element*
t: substance; ct: mixture, compound.

Here the entry shows that the chemical term *element* is used in relation to *substances* and is contrasted with *mixture, compound*. Thus substitution of *compound* in a context requiring *mixture* could lead to a disastrous translation.

Repeated use of the Thesaurus when dealing with translation assignments soon reveals the advantages of this dictionary arrangement. The compact structure may present an initial handicap to inexperienced translators, yet it soon becomes an asset. Moreover, it is in some ways more convenient for translators as it encourages systematic thinking, as opposed to blind scanning.

14.7 Synonymy

The Thesaurus employs descriptors for various types of synonym: *s*, *cs*, *ps*, *nps*. These correspond to *true*, *contextual*, *preferred* and *non-preferred synonym* respectively.

i. ***true synonym, acronym, abbreviation***

In practice, the number of concepts occuring in technology is so vast and the terminology available so restricted that "true" synonyms (terms which are "100% interchangeable") almost never occur. The Thesaurus uses the descriptor *s* mainly in cases where a short form of a technical expression is more common than the full term, for instance:

accelerator AUTO *Gaspedal*
s: accelerator pedal; ct: clutch, brake pedal.

It is unnecessary to include *accelerator pedal* as a separate thesaurus entry as it is easily locatable (via the browser command *Find*, etc.) and any additional information could be redundant.

The descriptor *s* is also used where the full form corresponds to the spoken form yet has been virtually completely replaced in the literature by an abbreviation or acronym. For example:

root-mean-square value ELEC *Effektivwert*
s: RMS value.

Here the Thesaurus supplies the appropriate definition not at the entry for the spoken form *root-mean-square value* but at the entry for the preferred form in the engineering literature, in this case at *RMS value*.

RMS value ELEC *Effektivwert*
m: wave, oscillation, ac signal; t: mean value.
RMS value MATH *Effektivwert*
d: square root of the mean of a set of squared values.

Other examples are the computing terms *ROM*, *RAM*, *CPU* and the nomenclature of semiconductor devices *fet*, *mosfet*, *ujt*.

ii. ***contextual synonym***
A more common category of synonym is the contextual synonym (*cs*), where the alternative term can be substituted in the majority of contexts, but not in every case. For instance:

V-belt AUTO *Keilriemen,-m*
cs: fan belt; u: pulley, engine, water pump, generator.
forward-bias mode ELNC *Durchlaßrichtung,-f*
cs: conducting mode; u: diode.

In most motor vehicles the *V-belt* (so-called because of its V-shape cross section) drives the *generator* and possibly the *water pump* to which a *cooling fan* is attached. It is unlikely in practice but engines are conceivable that are cooled entirely by *electric* as opposed to belt-driven fans. In the latter case, the substitution *fan belt* for *Keilriemen* would not be correct as the belt concerned would have a different main function (e.g driving the generator).

iii. ***preferred/non-preferred synonym***
The remaining categories *ps* (preferred synonym) and *nps* (non-preferred synonym) are small ones and occur mainly in connection with layman expressions and nearly obsolete terms. Layman terms may occur in technical texts and should be understood by the translator, but they should not be used in translation unless the context specifically requires it. Obsolete synonyms should not be used in translation at all. Thus, the entries:

coil *lmn.*	ELEC	*Spule,-f*
ps: inductor.		
collector *obs.*	ELEC	*Kollektor,-m*
ps: commutator; u: electric motor.		
spark plug	AUTO	*Zündkerze,-f*
nps: sparking plug.		

indicate that *coil* is a common layman term for what the circuit designer calls an *inductor*, and that the former term *collector* in the context of *electric motor* has been replaced by *commutator* throughout the electrical engineering industry; these terms are dying out and may not be understood in the sense intended by the future generation of engineers. The third example lists an alternative term *sparking plug*, which exists but whose usage is dwindling.

14.8 Other Terminological Associations

This section discusses the remaining descriptors: *d* (*designates*), *m* (*measurable parameter of*), *tu* (*typical unit*), *a* (*associated with*).

i. ***designation, description, definition***

Consider the entry below, taken from the field of semiconductor device technology:

extrinsic material	SEMI	*dotierter Halbleiter*
d: doped semiconductor material.		

The example implies that the term *extrinsic material* denotes a *semiconductor material* that has been subjected to *doping*. The descriptor *d* does not indicate a hierarchic or hyponymous relationship among terms. It refers to the *concept* denoted by the entry expression, and designates semantic attributes that distinguish the entry term from other related expressions, such as *intrinsic material*.

ii. ***metric parameter, typical unit***

Entries involving the descriptors *m, tu* concern engineering terms that correspond to *parameters*, or more specifically *physical quantities* (Chap. 1–2). Consider the examples:

inductance	ELEC	*Induktivität,-f*
m: inductor; tu: henry.		

inductance PHYS *Induktanz,-f*
s: electromagnetic inductance; u: induced emf.

The first entry helps the translator overcome the common pitfall of imagining that *inductance* is always expressed in German by *Induktanz*. It reveals that *inductance* is also a measurable parameter (*m*) characterising the electrical device *inductor*; if the context involves the typical unit (*tu*) *henry* then the translation of *inductance* by *Induktivität* is likely to be the correct one. The second entry provides the explanation for the translator's initial problem: namely that the term *inductance* has another meaning in English, associated with the first but with a broader significance and not necessarily to do with the electrical device *inductor*.

iii. ***general/contextual association***
Finally attention is drawn to the thesaurus descriptor *a* (*associated with*) corresponding to the semantic relation *associative* introduced in connection with hierarchic organisation. Some readers may have observed that this category is present already in the descriptor *u* (*used in connection with*). Why then should a second descriptor be introduced? The answer is purely for convenience.

For a term like *acid level* which is entered in the Thesaurus as follows:

acid level AUTO *Säurestand,-m*
a: battery.

there is indeed no good reason why the definition must be *associated with* (*a*) as opposed to *used in connection with* (*u*) a *battery*. In other cases, however, an alternative descriptor helps to distinguish the type or degree of association:

rocker shaft ENGN *Kipphebelbrücke,-f*
u: cylinder head; a: valves.
induced emf ELEC *Induktionsspannung,-f*
u: winding; a: transformer, generator.

The entries above indicate that the term *rocker shaft* is used in connection with the *cylinder head* of an *automobile engine* and is associated with the concept *valves* (inlet valves, exhaust valves), whereas *induced emf* is an electrical term used in connection with a *winding* and associated with a *transformer* or *generator*. The fact that two associative descriptors are employed makes the definition easier to read.

14.9 Morphology

Thesauri of this type have an inbuilt tendency to become rather large. Consequently every attempt was made to avoid repetition of information which is accessible in the TPD, especially where concepts are easily understandable and the reader merely requires the German equivalent. Thanks to the morphological properties of German it is possible to omit a few categories of terminology altogether. These are listed below:

i. Terms like *acoustics*, *thermodynamics*, *ultrasonics* which refer to broad subject areas are not defined in the Thesaurus. German equivalents are locatable in the Polyseme Dictionary at the entry "*-lehre*": *Schallehre*, *Wärmelehre*, etc.
ii. The same applies to terms like *coolant*, *detergent*, *lubricant*, everyday concepts refering to materials used in the household, motor vehicle, etc. whose German equivalents appear in the TPD under *-mittel*.
iii. And it applies to compounds involving *cutters*, *pincers*, *pliers*, *tongs*, *strippers*. These are located at the TPD entry *Zange*.

The Thesaurus attempts to clarify as many as possible of the engineering concepts appearing in the TPD by singling out their fundamental constituent terminology. Access to a large number of TPD compounds and to useful structured definitions of associated concepts in the Thesaurus is possible too, but detective skills and lexicological initiative are sometimes required, as so often in technical translation work.

Unit 15

Electrical Sciences

This unit begins on an odd note. The term *Electrical Sciences* is not standard. It serves merely as a *heading*, enabling the contents of two further chapters to be discussed and related to three earlier ones. It introduces Chapters 7 and 15, dealing with *Circuit Technology* and *Light/Heavy Electrical Engineering*, and relates them to Chapters 2, 5 and 6 which concern *Basic Electricity*, *Semiconductors* and *Electronics*. The unit thus ties up a few remaining loose ends necessary for a wide-angle view of the gigantic field of *Electrical Engineering*. Unit 16, headed *Mechanical Sciences*, serves a similar objective for the *Mechanical Engineering* disciplines.

For readers who feel uncomfortable in the electrical fields there are a number of disk illustrations to ease comprehension, in particular in the thumbnail sections:

i. Basic Electrical Engineering — *ac/dc, phase,* RMS *value, reactance*
ii. Electronic Circuit Design — *device characteristics, circuit diagrams*
iii. Electrical Engineering — *motor, generator (starter, dynamo, alternator)*

The unit itself focusses on general description rather than specific information, drawing attention to a number of contrasts, *electrics/electronics, mmf/emf, coil/inductor*, and subject-field inconsistencies, such as *power* (Ge. *Leistung, Energie, Strom*).

15.1 Circuit Design

The full list of standard electronic circuits is immense and the degree of complexity often bewildering for a non-technical person, but certain circuits reappear in different electrical appliances again and again. Chapter 7 describes those circuit types most frequently encountered in engineering applications, the fundamental operating mechanisms involved and the main electronic components concerned.

Discrete electronic components, such as *transistors, diodes, resistors, capacitors,* are interconnected in various configurations to form standard *circuit modules.* Three types are considered here: *amplifiers, multivibrators, oscillators.* Amplifiers increase the *amplitude* of small ac voltages, known as *signals,* without changing their frequency. Oscillators either generate or filter out signals of a particular frequency. Switching circuits respond to *pulsed signals* (i.e. sudden voltage surges; Ge. *Spannungsimpuls*) as opposed to *sinusoidal* ones (smooth regular voltage waveforms; Ge. *sinusförmige Signalwelle*) and involve elaborately interconnected arrangements of more fundamental switching-circuit modules known as *multivibrators.* Designs of amplifier and switching circuitry are centred on *transistor* applications, whereas those of oscillator circuitry are often based on parallel configurations of *capacitors* with *inductors.*

i. ***Multivibrator Circuitry***

Electronic switching circuits (Ge. *Schaltkreis*) underlie the operation of *flashing light systems,* such as the *indicator lights,* Am. *tail lamps,* of a motor vehicle. They are also used in conjunction with *sensors* for triggering fire alarms, and for a wide range of other everyday applications. Basic switching modules are known in the trade as *multivibrators* or occasionally by the layman term *flip-flop.* There are three types, *monostable, bistable* and *astable multivibrators,* distinguished according to whether the switching circuit has one stable output state (monostable), two (bistable), or whether it continually switches itself from one state to the other (astable).

Combinations of multivibrators lead to other modular units employed in digital systems and constitute the next order of switching circuit complexity:

clock timer	digital reference
logic gate	digital processing
register	digital counting
storage cell	digital memory

Logic gates are subdivided into *AND-, OR-, NAND-, NOR-* and *NOT-gates* according to the corresponding logic function. The other concepts may be vaguely familiar to the reader already. They all relate to the *hardware* of computer systems.

ii. ***Oscillator Circuitry***

Oscillator circuits (Ge. *Schwingkreis*) provide *voltage signals* of a predefined *constant frequency* and at *constant amplitude.* An obvious intuitive application of these circuits is the generation of the individual *key-notes* of an electronic organ or other keyboard instrument. In the first electronic organs good results were obtained by using so-called *LC oscillators*, different parallel arrangements of a specific *capacitor* with the appropriate *inductor*, which *resonate* at the frequencies required. The standard electronics symbols for inductance and capacitance are "L" and "C" (Chap. 2).

In an *LC oscillator*, electrical energy is transferred at a constant rate between the two devices, rather like the way the *spark energy* is transferred between the *condenser* (the ignition capacitor) and the *ignition coil* of an ordinary motor vehicle. Instead of the energy being suddenly released however, i.e. as a spark, it is continually transferred back and forth between the inductor and capacitor, thereby generating a *sinusoidal voltage wave* at a constant frequency which produces a synthesised musical note (Ge. *Ton*) of a keyboard or other instrument.

Similar results can be obtained, on a large scale, more cheaply *without* inductors by employing sophisticated electronic circuitry manufactured as *integrated circuits* (IC's). Waveforms can also be reproduced *digitally*, which is another approach to electronic keyboard design these days. LC oscillators are then used merely as *filters* in various applications.

iii. ***Amplifier Circuitry***
This is discussed in detail in the chapter, where distinctions such as *pre-amplifier/power amplifier, current/voltage amplifier, feedback system* are elaborated. There is also a detailed account of the devices employed, *thermionic tubes, diodes, transistors, thyristors*, etc., in amplifiers themselves and other electronic circuitry.

Other standard circuits, revealing international *symbols* relating to the range of electronic devices employed, appear in the *disk illustrations*. Those headed *Voltage Regulator, AC/DC Conversion* are directly relevant to the next section, the *power supply* of an electronic appliance.

15.2 Power Supply Unit

Every electronic circuit requires a *dc power supply* of a certain fixed voltage in order to operate or *bias* the respective components. The *circuit bias* may be obtained from *batteries* but usually it originates from the *ac mains*. Thus an electronic power supply does not really *supply* power at all. It simply converts the *ac mains voltage* into a *dc voltage* suitable for operating circuitry. The *power-*

supply unit (Ge. *Netzteil*) of any electronic equipment, including simple radios, cassette recorders and computers, is therefore itself a relatively sophisticated piece of circuitry, employing transformers, diodes, capacitors, transistors, IC's and other electronic devices.

A *power supply circuit* involves the following operations: *transformation, rectification, smoothing* and *regulation*:

i. The voltage level is set in the first stage: *transformation*. This operation implies the conversion of the ac mains voltage (220 V) into another ac voltage of the required level (e.g. 25 V). It requires a single circuit component, a suitable *transformer*.
ii. *Rectification* concerns the conversion of the smaller *sinusoidal ac voltage* into a *rectified ac voltage*, namely one in which the waveform varies sinusoidally but remains always *positive* (or always *negative*). Semiconductor diodes (or diode combinations) are used for this purpose.
iii. After the *rectifier* stage follows the *smoothing stage*, where the *rectified ac* voltage waveform is converted into *dc* by means of large capacitors.
iv. The result, however, is not a perfectly smooth constant voltage like that of a battery. It contains a small ripple. *Ripple voltages* (Ge. *Brummspannung*) are responsible for *mains hum*, a steady low-pitched whining sound (50 Hz) characterising many cheap amplifiers, record-players and very old radios from the thirties (so-called *wirelesses*).

A good power supply is one with a low *ripple factor*, that is to say one where the ratio of the ripple voltage to the dc output voltage is as low as possible, preferably less than 0.1%. To achieve this, relatively sophisticated circuitry is required, involving *zener diodes* and a *transistor feedback* system. A power supply providing a smooth, stable dc output regardless of load or temperature extremes is said to be *regulated* (Ge. *stabilisiert*).

15.3 Hierarchic Arrangement

Like other chapters, Chapter 7 has its own bilingual microthesaurus and a number of hierarchic glossaries defining relevant concepts discussed only briefly or not at all in the chapter itself, such as *tuner, suppressor, receiver*. There are many NCNs in the thesaurus, *circuitry, interference, reception, rectification*, and occasionally in the hierarchic arrangements. Figure 7B contains terminology relevant to the previous section. It is reproduced below:

1	power supply	*Stromversorgung f*
11g	mains source	*Netz n*
111m	mains voltage	*Netzspannung f*
112m	mains frequency	*Netzfrequenz f*
2	dc power supply	*Netzgerät n*
21a	mains input	*(Netzspannung)*
22a	dc output	*(Ausgangsspannung)*
221m	output voltage	*Ausgangsspannung*
222m	ripple voltage	*Brummspannung*
223m	internal resistance	*Innenwiderstand m*
3	voltage conversion stage (ac to dc)	
31g	transformation NCN	*Umspannung*
32g	rectification NCN	*Gleichrichtung*
33g	smoothing NCN	*Glättung*
34g	regulation NCN	*Stabilisierung*
35g	output stage	*Ausgangsstufe*

Hopefully, most readers are by now conversant with this lexicological technique (employed on the disk from Chapter 2 onwards). It is quite straightforward. Terms on the right constitute either exact translations of those on the left, or sensible translations (bracketed expressions) where an exact German description of the concept implied does not provide a sensible substitution:

mains input	*(Netzspannung)*
dc output	*(Ausgangsspannung)*

The relational descriptor *g* indicates types or subcategories of the superordinate concept, a *generic* relationship. Thus *mains source* is one type of *power supply* (in the broad sense), *transformation* is one *stage* involved in ac/dc voltage conversion. The descriptor *m* designates a *metric* relationship, indicating that *output voltage, ripple voltage, internal resistance* are *parameters* relating to the concept *(dc) output.*

Note: The orthographic conventions for abbreviations regarding the concepts *direct/alternating current* (i.e. *dc/ac, d.c./a.c., DC/AC, D.C./A.C.*) vary within technical language and differ from those of general language. The convention employed in this handbook and (to avoid confusion) on most of the disk, *dc/ac*, is that of electrical or electronic circuit design, but there is no need for translators to adopt the same convention automatically.

15.4 Light/Heavy Electrical Engineering

Chapter 15 begins with a brief description of the above disciplines, and draws attention to slight variations in *terminology* encountered by translators when moving from one electrical subfield to another (e.g. *Spannung*: voltage, tension, emf; *Kondensator*: capacitor, condenser; *Widerstand*: resistor, rheostat). These variations are not necessarily apparent in both languages simultaneously and may have important consequences on an individual linguist's translation proficiency. Later sections continue with the topic of Light Electrical Engineering (Ge. *Schwachstromtechnik*), and demonstrate distinctions between *ac* and *dc machines.* The final section extends to Heavy Electrical Engineering (Ge. *Starkstromtechnik*) with illustrations of the significance of *phasing* in connection with *power transmission*, and descriptions of *motors, generators* and other *electrical machinery* which operate from a *three-phase* as opposed to a *single-phase supply.*

This unit discusses an extract from these fields. But first, for demonstration purposes, a brief look at Figure 15A, a terminology of *electric motors*:

1	motor	*Elektromotor m*
11g	dc motor	*Gleichstrommotor*
12g	ac motor	*Wechselstrommotor*
13p	field winding(s)	*Feldspule f*
14p	commutator	*Kommutator m*
15p	carbon brushes	*Kohlebürsten pl*
16p	armature	*Anker m*
161m	armature speed	*Drehzahl f*
2	ac motor	*Wechselstrommotor*
21g	three-phase motor	*Drehstrommotor*
211g	synchronous motor	*Synchronmotor*
22a	three-phase mains	*Dreiphasennetz n*
3	field winding	*Feldspule f*
31a	magnetic flux	*magnetischer Fluß*
311m	flux density	*Flußdichte f*
312m	induced emf	*Induktionsspannung f*
313m	rate of change	*Änderungsgeschwindigkeit*

The hierarchic list reveals the usual *generic* relationships:

three-phase motor is a type of *ac motor*
synchronous motor is a type of *three-phase motor*

and also *partitive* ones:

commutator, field windings, brushes are parts of a *motor*

as well as *metric* relations:

flux density, induced emf are parameters relating to *magnetic flux*

The necessary mental substitution processes are similar to those employed for other hierarchic term lists or thesauri. The intellectual investment pays off, however, when the reader is faced with an urgent translation assignment and is desperately using the disk merely as a dictionary. In this case, a number of *visual aids* are available for comparison. They appear in the subsection *Electrical Engineering* of the disk illustrations.

15.5 Electrical/Electronic/Magnetic

Electrical Engineering is such a vast field that not even specialists themselves are familiar with the full range of associated terminology. Moreover, certain concepts have designations which differ slightly according to the connotations involved. For instance, the German concept *elektrische Spannung— voltage, emf* (*electromotive force*). Chapter 15 clears up a number of potential misunderstandings.

Electrical engineers work with *electrical machinery* involving powers up to the megawatt range, while electronic engineers deal with applications ranging from *assembly-line robotics* to *spacecraft control*, working mostly at the milli- or microwatt level. Electrical engineers work closely with *magnetic devices*, such as *inductors, transformers, field windings*. Electronics experts tend to avoid such devices as far as possible and have no need for expressions like *emf, mmf* (electromotive and magnetomotive force) which derive from analogies and parallels within the fields of Electricity and Magnetism.

Resistors used in Light Electrical Engineering often consist of flat strips of metal (copper, etc.) or *wire coils*, and look very different to their counterparts in Electronics, the latter being tiny solid devices manufactured from semiconductor materials, or substances other than metallic alloys. Electrical terminology tends to be older than that of Electronics: *capacitors* are still called *condensers* in a few areas; *inductors* are referred to as *chokes*; certain *wire-wound potentiometers* are termed *rheostats*. Translators are frequently confused where German uses the same terminology for both electrical and electronic devices: *Spule* (choke, inductor), *Kondensator* (condenser, capacitor), *Schiebewiderstand* (rheostat, sliding resistor).

On the point of *grammar*, non-native speakers should note that the adjectives *electric* and *electrical* are both possible in certain compounds, for example in combination with *power*, *energy*, *field*, *charge*, but in other compounds there is only one alternative: *electrical machinery; electric motor*, *electric dish-washer*, *electric guitar*. In the case of *electronic* or *magnetic*, however, the choice is much simpler. Alternatives with *-al* do not exist.

15.6 Electrics, Electronics

The terms *electrics* and *electronics* have little in common. *Electronics* is an *engineering* discipline, whereas *electrics* is a pragmatic abbreviation for *electrical equipment* and appears in compound terms where the full expression is a little clumsy: *auto-electrics*, *lathe electrics*, *household electrics*. Occasionally, however, this distinction is blurred. For example, a distinction between *auto-electrics* (dashboard, wipers, lighting, window winders) and *auto-electronics* (electronic ignition, electronic fuel injection) is now emerging.

The expression *household electrics* refers to the system of *insulated cables*, *mains sockets*, *mains switches*, *light switches*, *junctions*, etc., installed by an *electrician* in a household, office, school or factory, the so-called *wiring system* (Ge. *Verkabelung*). Each household electrical appliance has a *mains lead* (Ge. *Netzkabel*) to which a *mains plug* (Ge. *Stecker*) is attached. When the appliance is operated, the lead is inserted (*plugged*) into a *mains socket* (Ge. *Steckdose*). American English employs the terminology *power cord*, *power plug*, *power socket*; rather misleadingly for non-native speakers though, *power sockets* themselves are colloquially referred to as *plugs*, so that *plugs* are effectively *plugged* into *plugs*!

15.7 Plugs, Fuses, Cut-Outs

Household mains plugs may have just two *pins* which provide the connection to the *live* and *neutral mains*, the in-coming cables leading to the *power grid* (Ge.

Stromnetz) and eventually right back to the local *power station* (Ge. *Kraftwerk*). But most British plugs have three pins, the third being at *earth potential* (Am. *ground potential*). German plugs are often *earthed* (Am. *grounded*) but instead of a third pin a metallic spring clip at the sides of the plug provides the necessary contact.

British *3-pin plugs* are *fused*, that is they contain *fuses* (Ge. *Schmelzsicherung*) designed to *blow* when a certain current is exceeded, the most common being *3-amp fuses* (for the *lighting systems*) and *13-amp fuses* for the mains sockets. British mains sockets themselves are also *fused*, namely wired to fuses in the *fuse-box* adjacent to the *electricity meter*.

The basic meaning of *fuse* is a tiny section of delicate *wire* designed to break when the temperature of the wire due to the current conducted becomes excessive, but there are a number of extended meanings. The adjective *fused* can mean containing or attached to a fuse, as in *fused plug*, *fused socket*, but the participle *fused* implies that a fuse has blown, for instance: "*The lights have fused*". Thus expressions like *fused plug*, *fused socket* acquire a second interpretation. Small interchangeable fuses, consisting of *fuse wire* inside a transparent casing attached to convenient lateral metallic *terminals*, are employed in televisions, video recorders, hi-fi systems, etc. to avoid destruction of delicate circuitry. These fuses generally *blow* between 1 mA and 1 A.

German and American household wiring systems are not fused but employ instead a system of *cut-outs* (Ge. *Ausschalt-Sicherung*), small automatic switches operated electrically by *relays* or other electrical devices which *flip* the switch in the event of an *overload*. Cut-outs are employed in household appliances, when the appliance itself is less delicate or the possibility of death by *electrocution* is more remote. In such cases *push-button cut-outs* are common, which are depressed when the overload has been removed.

15.8 Power, Performance, Energy

Chapter 15 goes a long way beyond simple household applications of electricity. There are detailed sections on aspects of *Auto-Electrics* revealing the underlying features and constituents of *batteries*, *starters*, *generators*, *dynamos*, *alternators*. It also deals with *electrical machinery*, revealing new terminological contrasts, *emf/mmf*, *resistance/reluctance*, as well as the first *magnetic/electromagnetic* parameters and units: *flux*, *ampere-turns*, *weber*. The chapter ends with a study of *high-voltage power generation/transmission*, discussing *single/three-phase power systems*, *power transmission lines*, *grid voltage* and other specialised concepts which demonstrate features unique to this area. This section looks at one frequent translation error in this field. It involves the simple German expression *Energie*.

A complete beginner in technical translation, on encountering the German term *Leistung*, picks up a technical dictionary, finds the possible substitutions *performance*, *output*, *capacity*, and selects one at random. Probably the most frequent English equivalent is none of these. In this handbook, on most of the disk, and indeed throughout engineering, the German concept *Leistung* mainly corresponds to *power*, a physical quantity measured in the SI unit *watt*.

Heavy electrical or power systems engineering constitutes no exception but the German terminology is less systematic and the nearest translational equivalent of *power* in the reverse direction is sometimes *Energie*, for example: *power transmission*, *power supply* (Ge. *Energieübertragung*, *Energieversorgung*). In other cases, German employs *Strom* as in *Stromversorgung*, *Stromleitung*, *Stromkabel* (E. *power supply*, *power line*, *power cable*). English, on the other hand, uses the expression *power supply* in a number of slightly different ways, and these can have different German equivalents.

Similar considerations apply to the global expression *Leitung* (lead, cable, power line). The lengthy microthesaurus of Figure 15B casts some light upon these distinctions, but like with so many other aspects of technical translation, a proficient translator must remain on continual alert.

Unit 16

Mechanical Sciences

Mechanical engineers design and construct *engines*, *turbines*, *drive systems*, *lifting gear* and lots of other equipment for application in specific branches of technology or industry. *Mechanical Engineering* in the broad sense of the term covers a vast field, with the result that engineering students specialise early and those opting for Aeronautical or Construction Engineering generally switch to separate courses. The e-book too divides the subject into many separate domains.

Despite this, certain terminology is common throughout the field. The basis of Mechanical Engineering lies in the branch of Physics known as *Newtonian Mechanics* and employs the terminology of Chapter 1. Other aspects are to be found in the chapter on *Automotive Engineering* (Chapter 8), in Chapter 13 focussing on *Construction Engineering* and in Chapter 14, headed *Mechanical Engineering*, where the latter deals with *railway vehicles*, *aircraft*, *shipping* and other aspects of the field not covered elsewhere.

This unit introduces *Machine Technology*, a subfield of Mechanical Engineering responsible for the manufacture of *machine parts* and the construction of *machinery*. It goes on to describe various sciences which developed mainly in the twentieth century under the collective heading of *Mechanics*. Detailed information appears in two separate areas of the disk: Chapters 9 and 12. The heading *Mechanical Sciences* is introduced for the author's convenience; it is not a standard engineering term. It merely serves to give the Unit a name, and enables it to combine two chapters and two different aspects of Mechanical Engineering, not mentioned so far.

16.1 Machine Tools

Machine technology includes the design of *workshop machines* used as *tools* to manufacture components of other machines: *lathes*, *drills*, *grinding*, *broaching*, *planing* and *milling machines*, and so on. These should perhaps be called "tool machines" (cf. Ge. *Werkzeugmaschinen*) but the standard term is *machine tool*. The use of machine tools to manufacture intricate components of *machinery*, *engines*, *gear assemblies*, *linkages*, individually is termed *metal-working*. The *metal worker* begins with a piece of metal of the appropriate material and can produce *cogs*, *gears*, *screws*, *bolts*, *sleeves*, *carburettor jets* and indeed any *precision component* required in Mechanical Engineering with the aid of machine tools installed at various *benches* (Ge. *Werkbank*) in the workshop. Repeated identical manufacturing and assembly processes are now carried out by *industrial robots*, especially in the Automobile Industry.

Metal workers and indeed all mechanical engineers need to be familiar with material properties and require an exact knowledge of the appropriate parameters, such as *melting point*, *fatigue strength*, *maximum shear stress*. They require a number of practical skills too, including *brazing*, *welding*, *forging*, *joining*, using *hand tools* as well as *machine tools*. Figure 9A lists nouns describing the properties of engineering materials and Figure 9B contains some verbs summarising the various operations involved in metal-working. The names of machine tools often consist of simple compounds involving these terms, such as *broaching/slotting/grinding/reaming machine*, which are abbreviated to *broacher*, *slotter*, *grinder*, *reamer* in the layman language.

The term *machining* covers all *cutting operations* using machine tools. But there are many other technical verbs in this area, some no doubt familiar to the reader, *bore*, *drill*, *grind*, *plane*, *punch*, others less familiar: *broach*, *finish*, *groove*, *mill*, *seam*, *slot*, *square*. These appear together with their German equivalents in Figure 9B alongside other useful terminology from related areas, *bond*, *braze*, *cast*, *forge*, *rivet*, *weld*.

16.2 Machines, Engines

Mechanical engineers are employed to install engines and other machinery in ships, submarines, cranes, elevators, escalators, factories and factory workshops. The terminology of Automobile Engineering applies to *engine systems* in general, and Mechanics terms, such as *energy*, *power*, *work*, *torsion*, *momentum*, appear repeatedly in texts concerning the design of such systems. Translators encounter three main types of engine:

i. *petrol engines* (as in motor vehicles)
ii. *steam engines* (in power stations, or the early railway industry)
iii. *jet engines* (in air- and spacecraft)

Petrol engines are found in many different forms. The smaller variety appearing in *chain saws, lawn mowers, mopeds*, with just a single *spark plug*, tend to be called *motors* in British as well as American English. *Diesel engines* and the motors of certain oil-fired heating systems operate along similar lines but require *ignition systems* only for starting the engine, if at all. The same applies to engines fired by *gaseous fuel*, such as *methane*. Most engines involve *reciprocating motion*, in other words *pistons* moving along *cylinders* requiring *inlet valves, exhaust valves*, etc. The *wankel engine* employs *rotary motion* at the inlet and exhaust stages and dispenses with some of the above components. Another class of *rotary machine* of an entirely different kind are *turbines*, large rotating wheels fitted with *vanes* (Ge. *Schaufel*) turned by *fluid pressure* (water, steam, etc.). These are important in *power stations*. The *steam engines* of ocean liners or railway locomotives, and the *jet engines* of the aircraft and the aerospace industry (Chap. 14) also require the talents of skilled machine technologists.

Discrepancies between British and American English in the area of Machine Technology itself are minimal. But the distinctions between *engine* and *motor* vary however, and present occasional terminological problems for translators, though rarely conceptual misunderstandings. Usually in British English the term *motor* implies *electric motor*, and *engine* covers everything else — *diesel engine, steam engine, rocket engine*, etc. But exceptions exist, usually introduced via American English, such as *motorboat*.

In contrast to the electronics and computing industries, attempts to standardise terminologies in the various mechanical engineering disciplines have not been entirely successful. Discrepancies between British and American nomenclature exist in the automobile industry — *anti-roll bar/sway bar, antifreeze/defreezer, float chamber/float bowl, oil sump/oil pan* (Figure 8A); in the terminology of railway vehicles — *goods truck/freight wagon* (Ge. *Güterwagon*); in the area of household water systems — *pipe/tube, tap/faucet* (Ge. *Wasserleitung, Wasserhahn*); in the names of objects such as: *adaptor/adapter, mould/mold, vice/vise, socket wrench/box spanner, circlip/snap ring* (Ge. *Paßstück, Gießform, Schraubstock, Steckschlüssel, Sicherungsring*).

Note: The terms *machine* and *machinery* are applicable to *gearing, lever arrangements, rotating shafts* and *drive systems* but normally exclude *engines* and *turbines* themselves. There is also the extended meaning *electrical machine* encountered in the previous unit, which includes *motors.*

16.3 Drive Systems

Three types of *drive system* are employed in mechanical engineering: *belt drive, chain drive* and *gear transmission.* Belt drive normally implies *open-belt drive* (Ge. *offener Riementrieb*), where the belt passes smoothly over two *pulleys* rotating in the same direction, as opposed to *crossed-belt drive* (Ge. *gekreuzter Riementrieb*), where the belt crosses over in the middle and reverses the direction of rotation of the second pulley. Crossed-belt drive is generally less efficient (more *slip*) and is limited to applications where broken belts are easily accessible, for instance in *weaving looms* and other *textile machinery.* Belts themselves can be flat, but some have a V-shape cross-section designed to fit into the pulleys. *Pulleys* associated with *V-belts* (Ge. *Keilriemen*) are normally designated *Scheiben* in German, whereas those designed for flat belts tend to be called *Rollen.*

Chain drive is used where the *shafts* involved are too far apart for *gear transmission* and too close for *belt drive*, or when access to broken belts is difficult. The chain passes over two *sprocket wheels* (Ge. *Kettenrad*) as on a bicycle. *Gear transmission* involves direct contact between interlocking gears which always rotate in opposite directions. In the case of *spur gears* (Ge. *Stirnrad*) the *gear teeth* are arranged on a narrow cylindrical surface and the two shafts rotate at different speeds in the same plane. The teeth of *bevel gears* (Ge. *Kegelrad*), however, are arranged on a conical surface and the shafts rotate at right-angles to one another. *Spur gears* appear in the *gearbox* of a motor vehicle. *Bevel gears* appear in the *differential assembly* and switch the *drive* from the *engine* and *propellor shaft* to the plane necessary to turn the *roadwheels.*

16.4 Newtonian Mechanics

Newton himself had no conception of *machinery* and was by no means in any sense of the word a *mechanical engineer.* His field of research concerned the mechanical behaviour of objects in the world around him and in the skies above.

But Newton was a great mathematician as well as a scientist. He developed methods for computing *forces* in both *static* and *dynamic* systems and promoted the basis of what later became known as *infinitessimal calculus* (Ge. *Differential- bzw Integralrechnung*). Theoretical Mechanics is therefore really a branch of Mathematics dealing with the mechanical behaviour of certain idealised systems where, for instance, energy losses due to *friction*, *heat*, *sound*, or small additional considerations resulting from the respective *weights* of individual structural components are neglected, in order to establish *models* for practical mechanical systems. *Newtonian Mechanics* is the expression normally used when a contrast is made with other explanations of mechanical behaviour, for instance *Relativity*, a science which focusses more upon the *energies* involved in mechanical motion, or *Quantum Mechanics*, a subject based largely on *probability theory* and *statistics*. Elsewhere the accepted jargon designation is *Classical Mechanics*, where the expression *classical* refers to Newton and a few successful earlier scientists dating back to Archimedes.

Classical Mechanics is divided into three subdisciplines discussed in Chapter 1: Statics, Kinematics, Dynamics. Statics covers *forces* acting upon *bodies* in a state of equilibrium (bridges, cables, support walls) and underlies modern Construction Technology. Kinematics concerns the *relative motions* of mechanical bodies, e.g. *gears*, *cams*, *linkages*, *pulleys*, and is directly relevant to the design of *production machinery*, as well as *bicycles*, *winches* and *clockwork devices*; it constitutes an important aspect of the modern Robotics industry too. Dynamics deals with the production of rectilinear, circular and other motion by the production and application of dynamic forces. It underlies the technologies of both *reciprocating engines* (petrol or steam engines), with their associated *transmission systems*, and the *jet engines* of Aeronautical or Aerospace Technology.

The division of Mechanics itself into subdisciplines is no longer as important to modern technologists as it was to 19th-century scientists when the field was at an early stage; broadly speaking, *Construction Engineering* can be regarded as a practical extension of Statics, *Machine Technology* derives from Kinematics, and most other areas of Mechanical Engineering, e.g. *Automobile Technology*, *Railway Systems*, *Nautical/Aeronautical Engineering*, are based on Dynamics.

16.5 Solid-Body Mechanics

Solid-Body Mechanics is an extension of Newton's *Statics* and supplies the theoretical basis for *Construction Engineering*, the design of *bridges*, *roofs*, *buildings* and other *frame structures*. The objective is to compute *bending, twisting* and *shearing forces* at different points in the structure so that *stresses* in construction materials are known in advance and *rupture* is avoided. The field relies a lot on *geometric calculation* employing *force polygons* (Ge. *Krafteck*) and especially on a technique known as *superposition*, whereby problems concerning strangely asymmetric structures are resolved by adding similar inverse asymmetric structures to the design drawing until a symmetrical structure is obtained. *Force values* calculated from the composite structural diagram then have to be halved or divided by an appropriate amount to provide results for the original asymmetric structure.

Having established a *force model* for the structure, the engineer or designer examines the *stress/strain* behaviour at different critical points, and establishes whether *elastic* or *plastic deformation* is taking place (temporary or permanent alterations) when the structure is *assembled* and *loaded*, in other words when *tiles* are placed on the roof or vehicles are crossing the bridge. *Thermal strain* resulting from extreme temperature changes has to be considered as well. *Bending* and *shear stresses* are then estimated and, if necessary, the structure is either strengthened at weak points or completely redesigned. The effects of twisting, so-called *torsional stress* are also investigated and appropriate precautions taken to avoid *rupture* or *fracture* due to this phenomenon. Finally, in order to prevent the structure from vibrating uncontrollably under unfortunate conditions, the engineer devises a model to analyse its *resonance behaviour*.

The terminology of Solid-Body Mechanics and the main conceptions involved are discussed in Chapters 1 and 13: *moment/torque, tension/compression, stress/strain, torsion/bending*, etc.

16.6 Fluid Mechanics

Whereas Newtonian Mechanics provides correct answers to information requested on the motion of objects through vacuum (Ge. *materiefreier Raum*) it breaks down when the *resistance* or *drag* of the *medium* (air, water, etc.) has to be taken into account. Newton could have told us how long it takes for a brick to fall to the ground from the top of a cliff, but not a newspaper. Other physicists such as Bernoulli, Stokes, Poiseuille provided the theoretical basis for these calculations. *Fluid Mechanics* (Ge. *Strömungsmechanik*) concerns the behaviour of solid objects moving through *fluids*, where the term *fluid* is employed in the broad engineering sense and covers both *liquids* and *gases*. The field covers the flow of fluids *through* solid objects too, such as *pipes, tubes, open drains, canals.*

Moving fluids are considered to be composed of lamina known as *streams.* The mathematical or graphical representation of a stream of fluid particles moving at the same *velocity* is referred to as a *streamline.* It provides *velocity vectors* illustrating the motion of small objects within the stream, thereby providing information on the *rate* of *fluid flow* (Ge. *Strömungsgeschwindigkeit*) near solid surfaces (e.g. inside a pipe), and information for *shaping* objects to move effectively and efficiently with or against the fluid flow, a science known as *streamlining.*

Hence, some fluid engineers devise *pipelines* for *petroleum* or *natural gas,* others develop *weather balloons* for specific destinations, and others design *bodies* for cars or *fuselages* for aircraft. Such scientists estimate *pressures* at different points in the fluid and take account of the effects of *viscosity,* a phenomenon which results from the internal friction forces acting between the molecules of the fluid.

16.7 Quantum Mechanics

Although Fluid Mechanics employs conceptions from other branches of Physics and a level of Mathematics far beyond anything Newton could have coped with, it is nevertheless distantly related to Newtonian Mechanics. Quantum Mechanics bears no relationship whatsoever and employs an entirely different terminology.

Quantum Mechanics is branch of the mathematical discipline known as Probability Theory and Statistics, and is responsible for describing the behaviour of tiny particles carrying minute amounts of *energy, charge, momentum,* etc. It was realised in the nineteen-twenties that nature as it stands could not exist if *quantities* such as these were indivisible indefinitely. There had to be a fundamental unit: the *quantum.* For the basic unit of *electric charge,* the *charge quantum,* the obvious value to take was the so-called *electronic charge* (Ge. *Elementarladung*), the charge of the electron itself, which is numerically equal to the proton charge. For other parameters such as the *energy quantum* it was not so easy to determine the precise amount, but it became apparent that certain phenomena at the level of Molecular Physics, such as the escape of an electron from an atom with the emission of a *photon* (Ge. *Lichtquant*), could only take place if the permissible *levels* of *energy excitation* of an electron or similar particle conform to a particular model, the so-called *energy band model.*

This leads indirectly to a measure of the fundamental *energy quantum*, that carried by the smallest particle of luminous energy, the *photon*. But recent work in the area of Sub-Atomic Physics dealing with constituents more fundamental than even neutrons and protons themselves, so-called *quarks* (Chapter 4) may lead to re-evaluations in this field.

Though hotly contested at the time, even by leading scientists such as Einstein, Quantum Mechanics (like Newtonian Mechanics) nevertheless provides many correct answers. Without it the whole field of Semiconductor Electronics could not have existed, and this would have had enormous repercussions on the *entertainment* and *computing industries*.

16.8 Celestial Mechanics

Celestial Mechanics deals with the motions of *asteroids*, *comets*, *moons*, *planets*, *stars* and even *galaxies* relative to one another. Though Newton did not realise that stars also change their relative positions and was probably unaware that other galaxies apart from our own exist, his description of the fundamental laws underlying the motions of all *celestial bodies* (Ge. *Himmelskörper*) was one of the first great successes of his new science *Mechanics*.

Newton provided the key to other terrestrial phenomena. For example, secondary variations in the levels of high tides result from the fact that the moon's orbit is *elliptical* rather than *circular*, and that the gravitational attraction from the sun varies due to the earth's non-circular orbit. The combination of these forces led to catastrophic flooding in the Netherlands in the seventeenth century. Such events occur only very rarely, sometimes every few centuries, but they are predictable. Moreover, it is interesting to note that *friction* resulting from tidal movements *retards* the earth's rotation about its axis, making the *terrestrial day* progressively longer. There is evidence to suggest that 370 million years ago, before the appearance of the first dinosaurs, the earth's day was less than 22 hours.

Newton's ideas reflect the kind of relativistic thinking involved in the analysis of complex structural or kinematic systems. Because the moon revolves every 28 days instead of 24 hours the stars circle the *lunar skies* much more slowly. So does the sun. Only the earth seems to be in a fixed position. This observation led the American astronaut Alan Bean to remark: "If man had evolved on the moon he would have worshipped the earth and thought of it as a gigantic eye in the sky". Since then, many more moons in our solar system have been discovered, and the planets have been studied from many different perspectives. Newton's Laws account for their relative motions. They also still provide the theoretical and practical basis for the *gravitational propulsion* of spacecraft from one *celestial body* to another.

Two centuries after Newton, a brilliant young scientist working from first principles established an entirely new theory of the conceptions underlying *gravitation* and *planetary motion*. Albert Einstein's *General Relativity* showed that although Newtonian Mechanics provides a lot of correct answers, they are for the wrong reasons. Einstein's clear insight into the relationships between *matter* and *energy* provided a new explanation of planetary motion and accounted for certain unexplained phenomena governing the orbital motion of Mercury. At the present time, Stephen Hawking, a tiny crippled scientist barely able to move and virtually unable to speak without the aid of computer keyset is pushing the frontiers of knowledge even beyond Einstein's grasp. Ironically, the one area where Newtonian Mechanics could one day become redundant is the one for which Newton first proposed it: Celestial Mechanics.

None of this would have surprised Newton, who regarded his new science as a theoretical toy which accounted for the relationship between *gravity* and *gravitation*, for the physical effects of *friction* between solid surfaces, and the *motions* and *turbulences* in liquids. But the fact that Newtonian Mechanics provides the intellectual basis of vast industries ranging from *bicycles* to *motor vehicles* to *aerospace technology* would no doubt have surpassed even his wildest expectations.

16.9 Subject Fields

Though many branches of science and technology originate from Newtonian Mechanics the broad engineering conception *Mechanics* normally covers the fields *Classical* and *Solid-Body Mechanics* only. *Fluid Mechanics*, in view of its complexity, is felt to be a separate discipline; *Celestial Mechanics* employs the doctrines of Einstein in preference to those of Newton; *Quantum Mechanics* is entirely unrelated to Mechanical Engineering, its applications lying in Molecular Physics, Nucleonics, Materials Science and Semiconductor Technology.

Subject fields emerge gradually over the course of time, and the names attached to them by technologists are often non-systematic. English itself is by no means symmetrical in the use of labels for fields of the type *Mechanics*, *Electronics*, *Aeronautics*, *Nucleonics*. There are many connotational inconsistencies:

i. Mechanics is a purely *scientific/mathematical* discipline with applications in many technological areas, of which Mechanical Engineering is only one;
ii. Electronics concerns *engineering* fields like Circuit Design, Tube Design, Semiconductor Development, the term being merely an *abbreviation* of Electronic Engineering;
iii. Aeronautics is an applied practical and theoretical science providing information needed for aircraft design and their safe construction, as well as for *guided missiles* and other *unmanned* flying objects, whereas the counterpart Aeronautical Engineering concerns chiefly *pilot/passenger aircraft*;
iv. Nucleonics is a subfield of Nuclear Engineering concerned with interactions between *nuclear, ionised* and *radioactive particles*, as opposed to the broad engineering field which includes the design of *breeders* and *reactors*.

Thus *Mechanics* is a science and not a branch of engineering, whereas *Electronics* occupies the opposite role. *Aeronautics* and *Nucleonics* are either scientific fields in their own right or subfields of engineering disciplines according to the context. By the same token, *Electrostatics*, the study of *charged bodies*, is a science only, whereas Optics, the design of *telescopes, microscopes* and other *optical equipment* can be either a science or a technology.

One of the difficulties encountered by translators, even when they are using good dictionaries with entries accurately distinguished by convenient *field labels*, is that the labels themselves can be understood differently in different languages. In Volume Two, a separate introductory section examines this problem and its relevance to the large dictionaries of the e-book.

Unit 17

Technical Collocation Dictionary

Whereas the TPD and Thesaurus concentrate on specialised terminology in isolation the Technical Collocation Dictionary (TCD) tries to illustrate meaning by way of *usage*. It deals with general vocabulary in specialised contexts, collocations of technical nouns with specific adjectives, concordances of nouns and verbs, mathematical expressions, prepositional phrases and many other important aspects of translation. Used appropriately, the TCD can provide short-cuts to a wide range of expertise unattainable from conventional lexicographical arrangements.

The collocations are accessible in a number of ways. The first part of this unit deals with *access*, the other sections explain the *content* of the TCD and distinguish a variety of its functions.

17.1 Access

Supposing a translator requires an English expression for *senkrecht* in connection with a Mechanical Engineering assignment. He looks up the term in the TCD and discovers an example of the use of *senkrecht* complete with an associated verb *stehen* and preposition *auf*:

> senkrecht: PHYS
> *Die Zentripetalkraft muß auf der Kreisbahn senkrecht stehen.*
> The centripetal force must be perpendicular to the orbital path.

This is *direct access.*

The translator may wonder whether this particular rendering has anything to do with terms *Kreisbahn, Zentripetalkraft.* He looks up first *Bahn* and then *Kraft* which provide the following information:

> *Bahn: bringen, senkrecht, umlaufen*
> *Kraft: angleichen, angreifen, anziehen, einwirken, ebenso …*

These entries indicate other entries in the TCD at which polysemes like *Bahn* (path, orbit), *Kraft* (force, thrust, traction) reappear. This is *indirect access* and this type of entry is called a *cross-collocation* to distinguish it from *direct entries*, such as *senkrecht.*

Technical translation involves native speakers in a considerable amount of problem-solving regarding terminology and semantics, but the difficulties of *non-native speakers* are even more acute as they lack the same *general* awareness of prepositions, adverbs, adjectives and their relationships with specific nouns and verbs. The remaining sections demonstrate how a wide range of potential translation problems are resolvable, with the help of the TCD, by using similar accessing techniques.

17.2 Specialised Nouns in Context

Some of the polysemes occuring in Volume 1, the Thesaurus or the TPD, such as *Widerstand* (drag, impedance, resistance, resistor), reappear in the TCD. Readers can supplement their understanding of concepts listed elsewhere in the book by observing terminology in context. The information is provided by entries of the type:

> *Widerstand: acht, anschwingen, charakteristisch, durchfließen, gering, Größe, Querschnitt, schalten, sprechen, Stärke, wärmeleitend*

which point to other TCD entries (e.g. *acht, anschwingen, charakteristisch*) that reveal collocations of *Widerstand.* The term may occur either in isolation or in compounds such as:

Luftwiderstand	atmospheric drag
Vorwiderstand	ballast resistor
Lastwiderstand	load resistance

in contexts illustrating different interpretations of the root concept. Root terms may appear in any position in the compound: initial, medial or final. Thus the entries at *Gitter, Last* points to collocations of:

Belüftungsgitter	ventilation grid
Gitteratom	lattice atom
Kristallgitter	crystal lattice
belastbar	capable of withstanding a load of …
Lastgerade	load line

Lastwiderstand	load resistor
Überlastung	overload

and that for *Leiter* directs the reader to a second set of collocations at *Halbleiter*. For frequent engineering terms, such as *Strom* (E. *current*), there are many collocations with different connotations: *Gleichstrom*, *Wechselstrom*, *Schwachstrom*. Those involving the same basic meaning are accessible via cross-collocations, but direct entries are provided for any exceptions:

ein stetiger Strom von geladenen Teilchen
a continuous stream of charged particles

17.3 Long Technical Expressions

Important expressions which are too *long* for ordinary technical dictionaries, or too difficult to define, nevertheless appear in the TCD. The collocations of *Satz* reveal:

der Satz von der Erhaltung der Energie
the Law of Conservation of Energy
der Impulssatz
the Momentum Principle/the Law of Conservation of Momentum
der binomische Lehrsatz
the Binomial Theorem

and another entry accessed via *Widerstand* reveals the concept:

Temperaturbeiwert des elektrischen Widerstandes
thermal coefficient of resistance

17.4 Technical Verbs, Specialised Predicates

For a small number of technical verbs, a dictionary approach similar to that of the Thesaurus is at least conceivable:

schalten	to change gear (AUTO, u: driving)
leerlaufen	to idle (AUTO, u: engine)
umwandeln	to transmutate (NUCL, u: radioactive materials)

But it is less appropriate when the expressions concern general verbs (e.g. *aushalten, aufnehmen, ausbessern*) used in specific engineering contexts. In such cases, it is better to provide an example of usage, either as a short phrase:

Wärme ableiten	to dissipate heat (ELNC)
Energie aufnehmen	to absorb energy (PHYS)
hohe Spannungen aushalten	to withstand high stresses (MECH)
Lackabsplitterungen ausbessern	touch up chipped paintwork (AUTO)
in Reihe geschaltet	connected in series (ELEC)
parallel geschaltet	connected in parallel (ELEC)

or as a complete sentence. Both models appear in the TCD.

The sentence approach is essential for verbs like *durchbrennen, ausdehnen, überschlagen* whose interpretations are restricted to a small group of grammatical subjects:

Die Glühbirne könnte durchbrennen.	The bulb might blow.
Das Gas dehnt sich aus.	The gas expands.
Ein Funke überschlägt.	Arcing occurs.

and for homonymous or polysemous verbs:

eine Schraube anziehen	to tighten a bolt (AUTO)
einen Körper anziehen	to attract an object (PHYS)
einen Motor zerlegen	to dismantle an engine (AUTO)
eine Kraft zerlegen	to resolve a force (PHYS)

Hence, specialised verbs or general verbs with specific interpretations in particular technical contexts constitute an important part of the TCD. Semantically contrasting entries appear adjacently, a feature which identifies polysemous verbs immediately.

17.5 Pragmatics

Some verb entries enable the dictionary user to avoid even tiny errors in translation, such as the substitutions *auffüllen* (fill up), *austauschen* (exchange), in contexts where these interpretations are not appropriate:

ein Loch im Kristallgitter auffüllen — to fill a hole in the crystal lattice
einen defekten Teil austauschen — to change a defective component

and the TCD draws attention to translations which would be wrong on pragmatic grounds, such as the substitution *send* for *schicken* in the example below:

Man schickt die Teilchen durch ein elektrisches Feld
The particles are *passed* through an electric field.

The word "sent" would be understood but does not occur in the above context in normal technical literature.

17.6 Syntax, Prepositions

As well as enabling readers unfamiliar with the literature of technology to avoid mis-translations, the TCD has another useful function. It encourages foreign speakers to refrain from clumsy non-English expressions like *"to prevent that a window mists up", *"in the case of falling below a certain voltage", and acquire a feeling for essential syntactic transformations:

das Beschlagen einer Fensterscheibe verhindern
to prevent a window pane from misting up (fogging up)
beim Unterschreiten einer gewissen Spannung
when the voltage falls below a certain minimum level

In contrast to specialised nouns, which are accessible in the TCD only indirectly, technical verbs are accessed directly, and the entry terms, e.g. *schicken, beschlagen, unterschreiten,* characterise the main translation difficulty.

It is sometimes convenient, however, to consult *indirect* collocations in order to examine concordances between specialised nouns and specialised verbs, and how this affects *prepositions.* The entry:

Kraft: angreifen, einwirken, richten

reveals that three verbs frequently occur in association with the noun *Kraft.* Collocation sentences involving this noun are accessible at the entries *angreifen, einwirken, richten*:

Die drei Kräfte greifen in einem Punkt an.
The three forces act at a point.
Eine Auftriebskraft wirkt auf den Körper ein.
An upward thrust acts on the body.
Die Kraft richtet sich gegen den Mittelpunkt.
The force acts towards the centre.

It is evident here that the distinction in English between the three verbal concepts focusses not on the verbs themselves but on the prepositions: *act at, act on, act towards.* Close examination of *verbs* associated with basic engineering concepts (e.g. *Kraft, Spannung, Ladung*) draws attention to aspects of the engineering chapters which the reader may have absorbed only passively.

17.7 Specialised Adjectives/Adverbs

Just as certain verbs only appear in specific engineering areas (e.g. *transmutate*, NUCL) so there are technical adjectives which can be included in a thesaurus:

kurzschlußfest	
able to withstand short-circuiting	ELNC, u: electronic device
spaltbar	
fissile, fissionable	NUCL, u: radiosubstances
speicheraufwendig	
using lots of memory	DPS, u: computer program;
requiring large amounts of store	DPS, u: data

But, although a thesaurus arrangement conveys the precise *meaning*, the collocational approach gives a clearer indication of *usage*. Indeed, for some adjectives, examples of usage are virtually obligatory:

leistungslos	without dissipating any power
leistungsgleich	having the same power rating
lacklösend	dissolving paintwork

The reason is that the English expressions only describe the meaning of the German and are not intended as target-language translations at all. Convenient dictionary equivalents for these terms simply do not exist. Grammatical, syntactic and other adjustments are necessary *in context* since translations like:

leistungslos steuern	*"to control without dissipating power"
eine lacklösende Flüssigkeit	*"a dissolving paintwork fluid"

are totally unacceptable. The TCD confronts this problem by providing full sentences illustrating *possible* translations of technical adjectives in narrowly defined contexts.

As an additional bonus, the collocational approach provides information on adjectives which have no real translational equivalent in isolation, such as *spanend, spanlos, unwesentlich, durchlässig,* only in expressions like:

spanende Formung	machining
spanlose Formung	shaping
unwesentliche Einflüße	second-order effects
elektrisch durchlässig	allowing current to flow

The same applies to expressions, such as *leitend* (conducting), whose interpretations are modified by the context:

nicht leitend	non-conducting; non-conductive;
wärmeleitend verbunden	providing good thermal conduction;
leitend mit Masse verbunden	directly earthed.

Moreover, whereas conventional dictionaries provide neat translations like:

entgegengesetzte Kräfte	opposite forces
entgegengesetzte Ladungen	opposite charges

they often fail to differentiate subtle distinctions in the meanings of adjectives:

entgegengesetzte Kräfte	forces acting in opposite directions
entgegengesetzte Ladungen	charges of opposite polarity

The TCD overcomes the problem by providing location facilities for other occurences of terms like *entgegengesetzt* and for other contexts involving the associated nouns (e.g. *Kraft, Ladung, Masse*).

17.8 Polysemous Adjectives/Adverbs

In the same way that general verbs appear in technical situations, so there are adjectives and adverbs which acquire different interpretations in specific engineering contexts. German makes extensive use of words like *groß, leicht, stark* which are a headache for translators:

eine große Wechselwirkung	a vigorous interaction
zwei gleich große Kräfte	two forces of equal magnitude
leicht aufeinander gleiten	slide easily over one another
leicht entflammbar	highly flammable
leicht spaltbar	extremely fissile
stark einschränken	greatly restrict
eine starke Überlastung	a serious overload

The TCD contains numerous direct or indirect collocations of these and other problematic adjectives frequently encountered in technology. There are troublesome adverbs too, *schwer*, *schwach*:

schwer brennbar	virtually non-flammable
schwach gebunden	loosely bonded

where the substitutions *lightly*, *heavily*, *weakly* are not possible in the engineering context concerned (Materials Science). Expressions like *gleich groß* require even closer attention to context:

of the same dimensions	machine components
of the same size	objects, storage tanks
of equal amplitude	waveforms, oscillations
of equal magnitude	vector, phasor or tensor parameters
equal (in value)	other parameters, mathematical quantities

Combinations of polysemous *adverbs/adjectives*, such as *gleich groß*, are sometimes exceptionally difficult to translate without background knowledge of the subject matter.

17.9 General Nouns, Specific Interpretation

The main purpose of the TCD is to illustrate the usage of general vocabulary in specific technical contexts. This largely implies verbs, adjectives, adverbs and prepositions, but there are also German *nouns* with specific context-dependent technical equivalents in English.

Den Strahl der Stroboskoplampe auf die Scheibe richten.
Shine the beam of the strobe lamp onto the pulley.
Das Auto mit einem Wasserstrahl klarspülen.
Rinse the vehicle using a jet of water.

Other examples are: *Größe* (amount, value, quantity, magnitude), *Weg* (path, distance), *Teil* (part, proportion), *Mittel* (mean, average). Unlike the facilities for specialised terms (e.g. *Widerstand*, *Leitung*, *Strom*) collocations of general nouns are entered adjacently in the TCD and are accessible directly. Here the indication of polysemy is the main object, illustration of usage a secondary one.

Strictly speaking, these nouns are neither *general* nor *technical* but some category in between. They are not restricted to a specific field, but nonetheless constitute a fundamental integral component of technical language itself.

17.10 Opposites, Contrasts

One feature of the TCD which has not yet been mentioned is *contrast*, namely whether *opposite meanings* of particular verbs and adjectives (antonyms, etc.) are listed in the dictionary and, if so, how they are located.

The existence of contrasting vocabulary is indicated by the thesaurus descriptor “ct” (contrasted with):

> nachlaufen: *ELEC* (ct: voreilen)
> *Der Strom läuft der Spannung um 90° nach.*
> The current lags the voltage by 90°.
>
> voreilen: *ELEC* (ct: nachlaufen)
> *Der Strom eilt der Spannung um 90° vor.*
> The current leads the voltage by 90°.

The above entries occur at different places in the dictionary but are nevertheless linked. The translator observes that the electrical terms *nachlaufen* (lag), *voreilen* (lead) are opposites and notices at the same time the correct prepositions. Occasionally, the contrasting expressions appear side by side:

> vorklappen: *AUTO*
> *ohne vorgeklappte Rücksitze (ct: mit vorgeklappten Rücksitzen)*
> with the rear seats erect (ct: with the rear seats folded forward)

17.11 Mathematics Expressions

Non-native English speakers may have difficulties determining the correct translation of *mit, bis zu* in expressions such as the following:

> *mit einer Geschwindigkeit bis zu 90% der Schallgeschwindigkeit*
> at a velocity approaching 90% of the speed of sound (90% Mach1)

But selection of the correct prepositional phrase is difficult for native speakers too in contexts where the main terminology derives from a subfield which is unfamiliar, such as Mathematics:

> *Leistung ist der Differentialquotient der Arbeit nach der Zeit.*
> Power is the differential coefficient of work with respect to time.
>
> *Arbeit ist das Wegintegral der Kraft.*
> Work is the integral of force with respect to distance.

Terms like *Differentialquotient*, *Integral* may mean little to translators but they are very common in engineering texts. Without suitable examples it takes a lot of time for translators to determine the correct prepositional expression "with respect to" associated with these terms. Special entry slots are reserved in the TCD for *mathematical expressions*. They are accessed directly by mathematical terms like *Quotient*, *Quadrat*, *Potenz*:

der Quotient aus der Spannung V und dem Strom I (d.h. V/I)
the ratio of the voltage V to the current I (i.e. V/I)
das Quadrat des elektrischen Stromes (I^2)
the square of the current
die dritte Potenz der Kupfermenge (m^3)
the mass of copper to the power 3
eine Zehnerpotenz niedriger als die Blindleistung
an order of magnitude lower than the reactive volt-amperage

Like the TPD and Thesaurus, the TCD is not a "dictionary" in the ordinary sense at all, but an organised didactic guide covering a variety of linguistic problems simultaneously. It reveals to both native and non-native English-speakers alike some of the most difficult aspects of technical translation. Other books concentrate on the phenomenon of translation errors. This book helps the reader to avoid them. Intelligent browsing of the TCD in conjunction with the other dictionaries will provide readers with the competence necessary to tackle professional technical translation assignments. Systematic organisation of their own *personal* data bases (terminology collections and collocations) will provide the necessary frame of mind.

Unit 18

Computer Engineering

Computer Engineering has a number of unique properties from the translation aspect. No other technical field has ever grown at such a phenomenal rate, but thanks to continual efforts from within the field to standardise its terminology, it can safely be assumed that each term corresponds to the same concept throughout the English-speaking world. Differences between British and American hardly exist. Many years ago, leading technologists working on opposite sides of the Atlantic agreed to adopt uniform nomenclature: *program*, *disk*, *dialog box*, *store device*. Not even the spelling differs. And there is a tendency for British specialists to extend this unusual trend of conformity to relevant, everyday terminology, such as *power connector*, *power cord*, *power outlet*, as opposed to *mains connector*, *mains lead*, *power socket* (Ge. *Netzanschluß*, *Netzkabel*, *Steckdose*). Some terms are accepted into German too, virtually without modification — *Cursor*, *Debugging*, *Mausklick*, *Monitor*, *RAM*, *Shortcut*, and the German layman language abounds with jargon, such as *ge-saved*, *booten*, *ausgelogged*.

Translators are not likely to make too many terminological errors in this field, except where dictionary entries are misleading because their terms refer to obsolete equipment, or have acquired a new significance. Chapter 11 provides a broad introduction to the field, outlining its history and highlighting problem areas for translators. Figures 11A–H provide a selection of microglossaries and thesauri for individual areas of the main field.

18.1 Dictionary Compilation

In no other discipline are the short-comings of the conventional printed dictionary more evident. The rate of change of technology in the area of Computer Systems is so phenomenal that printed dictionaries can never hope to contain a complete up-to-the-minute terminology of the subject. Fortunately, computer manufacturers themselves have been aware of the problem for a long time.

German institutions like the *Bundessprachenamt* and the translation department of the *Siemens* enterprise led the way to term banks in the nineteen-seventies, but even though these future-orientated institutions took the sensible precaution of storing their terminology with appropriate *"classification labels"* (Ge. *Sachgebietsschlüssel*), tantamount to the *Field Codes* of the dictionaries of Volume 2, the labels proved to be inadequate for subsequent purposes of *term deletion*. Complete terminologies from the nineteen-seventies, in areas such as *punch-card hoppers, magnetic core store, magnetic cards, magnetic tape decks, paper-tape readers*, are only of academic interest and should no longer be present in any published dictionaries. The same applies to early home computers, with terms like *tape drive, modulator, matrix printer.*

To the lexicographer working in this area, the erasure of obsolete terminology constitutes just as much a problem as the specification and compilation of new technical expressions. This applies to *electronic* as well as *printed* dictionaries. Hence, readers hoping to find large, elegantly structured, hierarchic term lists and thesauri, similar to those of other chapters will be disappointed. The terminology of Computer Engineering is too unstable to warrant this, except on a small or temporary scale, as the field is in a perpetual state of transition. Instead, the glossaries of the chapter focus upon:

i. a few areas where terminology has remained stable for many years and where German shows no current tendency to adopt English expressions face-value: *keyboards, character terminology, text-editing, circuit boards.*
ii. hardware conceptions common to industries centred around the main field, such as *CD writers, scanners, photoprinters*;
iii. the terminologies of certain common software systems, such as the document processor *Word.*

18.2 Logic Gates, Memory Modules

The terminology of hardware *circuitry* derives from Semiconductor Electronics, an area responsible for both *logic* and *memory modules.* Elementary *logic circuits* consist of *gates*, among which are *AND-, OR-, NAND-, NOR-* and *NOT-gates* (negation). *Logic gates* carry out logic operations on the binary data of *memory cells.* These operations are similar to those involved in the formal solution of philosophical arguments, where logic manipulations are performed on simple semantic propositions. The output states (true/false, on/off, 0/1) are *functions* of the *input states* and equivalent to the philosopher's *truth table* entries.

Logic gates consist of tiny electronic switching circuits, so-called *astable multivibrators*, whose output corresponds to one of two stable states. The layman term for this and other types of multivibrator, namely *flipflop* has also made its way into German. *Flipflops* are combined to form the *binary counters, decoders, registers* and *adders* required to carry out the fundamental *logic* and *arithmetic* operations involved in *electronic data-processing.*

Memory modules consist of *cells* divided into *bytes* and ultimately *bits.* The *bit* (derived from *binary digit*) is the smallest fragment of binary information and corresponds to one of two states (0/1, ON/OFF, TRUE/FALSE, etc.) of an element of digital hardware. It can be regarded as a single *switch.* A *byte* consists of eight adjacent *bits*; since the latter can each have two states, one byte provides 256 different combinations (2 to the power 8). The series of adjacent bytes constituting one *memory cell* is termed the *machine word*, or simply *word.* The engineering concept has no direct relationship to the natural language *word.*

Early flipflops employed first *relays* and later *transistors* as the main *switching devices.* Subsequently *logic gates* were manufactured as *IC's*, followed by *counters, decoders, registers* and *memory units* in *modular form.* Complete *processors* and *memory units* are now available as *IC-modules*, such as the *CPU* and *ROM* units of the common *PC.* The gradual evolution of flipflops and other devices led to the various different *computer generations* ranging from early *tube computers* to the modern *PC.* The first advances were dependent on *hardware* (electronics, etc.) but recent computer generations seem to be just as dependent on *software* fields, particularly *information processing* and *artificial intelligence.*

18.3 Microminiaturisation

By virtue of the thousands or even millions of tiny identical units involved in the design of *logic circuits, registers, memory modules*, Computer Engineering is a field where *microminiaturisation* has been taken to extremes. Indeed, circuit design using *discrete components* (Chap. 6–7) is only practicable at this level on a limited scale. Circuit-testing on a large scale, involving multiple interconnection of different *modules*, is only possible by *synthesis*: using *mathematical models* in computers themselves. As well as assisting in the design of their own ICs, computers (robotic processors) also manufacture them. Repairs or amendments to computer systems, such as the insertion of a *graphics card* or a *sound board*, merely involve slotting mass-produced ready-made units into an existing structure. *Skill* is needed more for the *software* aspect, namely *configuration* of the system so that the computer *knows* that certain operations have been carried out on its anatomy and can provide access to the units concerned.

Ordinary Electrical Engineering provides a section of the technical terminology. Most computers contain *fans* to prevent their intricate components from overheating. They also have *monitors* with *screens* and the usual range of *knobs* found on a TV set — *brightness, contrast, horizontal/vertical hold*, though not in the same places. But these too are rapidly disappearing. Modern monitors contain a small central processor of their own, which enables screen settings to be carried out via the mouse. A personal computer with separate *disk drive, monitor* and *printer* has several *mains leads* and various *connection cables* which *slot* into the appropriate *sockets*, though here translators may need to readjust their terminology. A computer technologist would speak of *power cords* and *data cables* that *plug* into the appropriate *ports.*

18.4 Processor, Calculator

A *processor* is a computational system which runs on a *fixed program.* The instructions cannot be changed by the user, although normally the fixed program itself can be substituted. Processors are used on *production lines* of various industries, including the electronics industry itself. *Industrial robots* found in car plants, etc. contain processors guided by *optical sensors.* A new generation of robots is now emerging with limited amounts of *intelligence* which enables them to memorise *motions* and in some cases to take decisions on the basis of recorded data. Their *limbs* are guided through the motion of a particular operation and the robot *learns interactively*, for instance, how to spot-weld the chassis of vehicles moving along an assembly line, how to interconnect cables, fit headlights, etc. Other robots can crawl along and inspect gas pipes, defuse bombs, carry out dangerous maintenance work at nuclear power stations or even bring meals to hospital patients. These are no longer *processors* in the strict sense of the term, but receive continual instructions, often from a *remote operator.* They relay back *optical* and other information (*temperature, radiation levels*, etc.) registered by their various *sensors.* Among other applications, robots have explored hidden tunnels in Egyptian pyramids and examined the surfaces of planets.

Calculators are simple pocket devices for carrying out arithmetic or other mathematical calculations. They too are *processors* but, unlike *robots*, the *data* is provided via the *keyset* and not by *sensors*. Readers will frequently encounter the expression *Rechner* in German computer manuals, journals, etc. To translate this as *calculator* would of course be nonsense. The expression is merely a stylistic or contextual equivalent of *Computer*.

Note: The term *data-processing system* (Ge. *elektronische Datenverarbeitungsanlage, EDV*) covers computers and processors and includes both hardware and software. But it is too clumsy and too similar to *data processor* for repeated use. It is generally avoided in English in favour of *computer system* or simply *computer*.

18.5 Disk, Memory, Store

A recent shift in the semantic significance of the unqualified English term *disk* has taken place. It corresponds to both *Festplatte* and *Diskette* in German. English distinguishes the concepts by employing the expression *hard disk* for the former. The original counterpart terms of previous decades, i.e. *floppy disk*, *diskette*, seem to be dying out. But the amount of data stored on (floppy) disk is minimal compared to that containable on *compact disk (CD)*. Thus it is possible that floppy disks themselves will disappear soon, and that the term *disk* will shift its significance a stage further.

The expressions *store* and *memory* (Ge. *Speicher*) are not synonymous in the field of Data Processing. *Store* is employed in the context of *permanent storage*, storage on a *medium* outside the *CPU*, such as *hard disk*, *disk*, *CD*, whereas *memory* implies storage within the central processing system itself, information which is immediately lost when the computer is switched off. The distinction only applies to the nouns *store/memory* and a few associated expressions such as *store/memory device*, *peripheral store/memory* (Ge. *Speichereinheit*, *Peripheriespeicher*). It does not extend to the equivalent verbs. There is no technical verb *memorize*, only *store*.

Note: At the time of writing, a new generation of software writers seem to be reverting to the former British spelling *disc*, which until recently had virtually disappeared. It is unlikely that these writers are British, as other anticipated deviations, such as *dialogue box*, *programme*, do not occur. It could be that they originate from India or another Asian country. Or perhaps the technical language of *software engineering* is simply making a small departure from that of *hardware* and from that of previous years.

18.6 Card, Board, Slot, Bus

The very early domestic computers offered virtually zero compatibility. Monitors were restricted to single computers, as were printers, and even floppy disks (diskettes) were formatted differently. But soon manufacturers were obliged to conform to market forces. The company *Amstrad* began offering disk-formatting facilities which were "IBM-compatible". Other producers followed suit, or went bankrupt. With better facilities for data-interchange, the *interfaces* (Ge. *Schnittstelle*) to the various *peripherals* (keyboard, mouse, monitor, joystick, etc.) were standardised, and more consistent software conceptions began to emerge. The situation has now gone full cycle. Expressions like *SCSI* (Small Computer System Interface), *AGP* (Accelerated Graphics Port) and *ISA* (Industry Standard Architecture) are becoming household words, in German too.

The wide variety of modern peripheral equipment, ranging from *scanners, synthesizers, photoprinters* and *modems* on the one hand, to *back-up drives, CD-writers, graphics/sound cards* on the other, mean that only small sections of a computer system need changing or *updating* at any one time. One recent innovation in interface technology, the *USB* (Universal Serial Bus) enables a phenomenal total of 127 peripheral devices to be interconnected and interlinked. They plug into one another in a kind of family tree structure. Despite the higgledy piggledy appearance of such systems, there are advantages which, until recently, to some engineers would have seemed astonishing. The various peripherals not only share the same data links (Ge. *Datenleitungen*) but also the same *power input* as the *main drive* and *mainboard* (Ge. *Netzeingang, Laufwerk, Hauptplatine*). Furthermore, the USB provides a facility known as *hot plug-in*: the peripherals can be plugged in and disconnected at will, while the computer is running and without *rebooting* it.

The computer industry now offers an extensive variety of *circuit modules, boards* and so-called *cards* (Ge. *Steckkarte*), which are easily installed by amateur enthusiasts at vacant prearranged locations inside their PC, so-called *expansion slots* (Ge. *Steckplatz*). The cards are connected to *bus slots* which direct data automatically along special cables, known as a *data lines* or *buses*, to the mainboard, peripheral boards (sound, graphics, etc.) and to the external *ports* for other peripheral equipment. The *housing* of virtually every modern computer, the metallic box containing the bare *slots* for the mainboard, CPU, hard disk and other electronic circuitry, conforms to a single universal design model (ISA design), providing a skeletal arrangement which enables *add-on boards/cards* to be slotted in and out almost at will.

Many companies are discovering that *peripherals technology* is a lucrative business. Figure 11H contains a useful introductory thesaurus to the area.

18.7 Works, Windows, Word

Just as MS-DOS (Microsoft Disk Operating System) enabled computers of the 1980s to achieve eventual compatibility, so one of the great innovations of the closing decade of the last millennium was *Microsoft Works*, which evolved into *Microsoft Office* and became familiar under the colloquial name of *Microsoft Windows*. There is little work for freelance translators in this area (if any), the various manuals and dialog facilities (*dialog boxes*, *Help windows*, etc.) being translated internally by professional software employees. Yet, despite the fact that translation is strictly controlled by a small number of individuals, 1:1 compatibility between languages has not been achieved by any means. The disk chapter takes a superficial glance at translation anomalies that have already occured in connection with various versions of the well-known program *Microsoft Word* (Word 97 onwards). It compares the situation to other areas of technology where similar anomalies occur and produce drastic repercussions for technical translators.

Most readers are familiar with the terminology of this section in one language already. It is summarised in Figures 11E–G. As a bonus, these glossaries provide rapid assistance to any reader suddenly forced to work with the same Microsoft system in the opposite language. But, other than that, the material can be taken lightly. It does not involve terminology that should be *permanently* at the fingertips of all translators. Each Microsoft application tends to be updated about every two years and, as the system changes, so too does the official translated version, as well as the terminologies used in computer manuals and by journals.

18.8 Word Terms

Cut and *Paste* are among the first *Word* commands mastered by any user. They are applicable to a variety of tasks. Photos or slides can be *pasted* into *documents*, as can musical or video arrangements, or information *downloaded* from the Internet. There is a sharp distinction in English between the commands *Paste* and *Insert*, the former implying data transferred via the so-called *clipboard* (Ge. *Zwischenablage*) and the latter — data input by other means (*keyboard*, etc.). German uses the expression *Einfügen* for both commands.

Similarly, there are technical verbs, such as *select*, *click*, *toggle*, that have fairly literal German translations: *auswählen*, *klicken*, *einschalten*. Others involve a slight semantic shift, e.g. *activate* (Ge. *aufrufen*). But a handful, for instance *scroll*, *align*, *justify*, seem impossible for German-speakers to handle, and oblige software translators to almost paraphrase. Consider, for example, the menu options *Decrease Indent*, *Justify*, *Align Left* (Ge. *linker Einzug*, *Blocksatz*, *linksbündig*). One of the repercussions from this difficulty is that neat clusters of English terminology, e.g. *scroll bar*, *scroll arrow*, *scroll block*, can become rather unwieldy and seemingly less inter-related in German: *Bildlaufleiste*, *Bildlaufpfeil*, *Rollbalken*. Verbs themselves are not the problem, as the opposite phenomenon also occurs, where English nouns become German *verbal* commands: *Bullets*, *Case*, *Page Setup* (Ge. *Aufzählung*, *Gross/Kleinschreibung*, *Seite Einrichten*).

And there are other inconsistencies. Terms like *menu bar*, *status bar*, *toolbar*, *taskbar*, *scroll bar* are familiar to Microsoft Word users in English-speaking countries. German adapts some of these expressions directly, e.g. *Menüleiste*, *Statusleiste*, *Taskleiste*, but has difficulty choosing a neat expression for *toolbar*, possibly because the *Tools* menu in German is named *Extras*. German employs a similar semantic shift in its choice of *Feld* as an equivalent expression to *box*, as in *list box*, *check box*, *dialog box* (Ge. *Listen-/Kontroll-/Dialogfeld*), which inadvertently destroys any possibility of overlap with the closely related concept *button*. A broad variety of concepts, such as *toolbar button*, *option*, *option box*, *menu item*, have therefore become subsumed under a global German concept *Schaltfläche*, for which there is no exact English equivalent. This in turn indirectly affects other expressions, e.g. *icon* (Ge. *Schaltflächensymbol*), which, though rapidly reduced to *Symbol*, gives a slightly different interpretation to *Standard Toolbar* (Ge. *Standard-Symbolleiste*). It is possible that Microsoft translators will eventually coax German users to accept expressions like *Tools*, *Toolbar*, *Ikon*, *Box*, *Button* into their language, just as *Menü*, *Option*, *Task* have been assimilated. But it is also possible that the writers will tighten up the English terminology so that fewer overlaps occur. The terminology, like the software, is in a state of flux. Anything can happen.

Just as each generation of software writers builds upon the work of the preceding one, so official translators are constrained by the terminologies of earlier versions. They may endeavour to *coax* the foreign language into a state of 1:1 translational equivalence, but they cannot impose it. It is not that the early translators did not understand the subject matter. They were simply hampered by the language itself and chose intelligent solutions at the time. It seems that technical terminology in other fields too is subjectable to a permanent tug-of-war, which is aggravated by the shortage of linguists with competent professional expertise, but which exists even in the absence of this shortage.

18.9 Polysemy, Polyonymy

Chapter 11 begins with a discussion of the practical and didactic benefits of *Computing* as a central topic for the training of translators, and compares it to *Automobile Engineering*. Neither area is particularly suitable for translator training, since they constitute opposite ends of the technical translation spectrum. But they provide interesting contrasts.

Polyonymy is a serious problem for the translation of automobile texts. There are major differences between British and American terminology, variations between different automobile companies, and substantial terminological discrepancies between this field and other important areas of engineering, e.g. *condenser, coil, high tension*. Similar situations occur in other areas of specialised translation too, such as Law, Economics, Politics, but it is very *untypical* of Engineering in general. By contrast, in the context of Computer Engineering, the problem of *polyonymy* is virtually absent. There are reasons for this.

Automobile Engineering is a market-orientated field which has arisen over a long period from the activities of individual private companies with little or no interest in terminology standardisation among themselves. Computer engineers, on the other hand, have been obliged by market pressures to make their products universally compatible, which has led indirectly to a rapidly expanding but universally acceptable terminology associated with these products. Most major engineering areas, such as Chemical, Mechanical, Nuclear or Electrical Engineering lie somewhere in the middle of these two extremes.

Though the *hardware* aspect of Computer Engineering provides suitable material for translator training, it seems that because of the way the industry itself has evolved, the *software* aspect provides useful practical experience only *from* English into another language. Getting university students to translate software instructions *into* English is rather like asking contemporaries of Martin Luther to translate the Bible from German back into Latin.

Nonetheless, the vast quantities of new software emerging each year and the ever-increasing variety of peripheral equipment and gadgets with digital operation (cameras, DVD, etc.) has already persuaded some European universities to offer translation courses for the field of Data Systems (Ge. *EDV*) as an alternative option separate from Engineering (Ge. *Technik*). In such courses, translation practice into English is necessary to ensure consistency in the university curricula, and to meet the demands of those software institutions whose organisational language happens to be other than English.

UNIT 19

Error Analysis

Most areas of knowledge, scientific, technological or otherwise, expand along a broad front in many different directions simultaneously. Linguistics tends to move in one direction at a time, *behaviourism*, *transformational grammar*, *comparative linguistics*, *text typology*, *translation theory*, with university teachers switching their enthusiasm about once every decade like flocks of birds. Any brilliant advance made nowadays in areas like *quantitative linguistics*, *case grammar*, *semantic syntax* might well be ignored because these areas no longer greatly interest those who rule in the academic world. Indeed some translation theorists divorce their area of specialisation from Linguistics altogether. By and large, *general semantics* has withstood the test of time and this is one area drawn heavily upon in the lexicological sections of the book. But perhaps too heavily for some theorists. This unit attempts to redress the balance and looks at technical translation not from the viewpoint of the translator but from that of the university language teacher.

19.1 Quality Assessment

Authorities on translator training differentiate between two extreme *views* common in language teaching: the *foreign language teacher's view* and the *professional translator's view* (Kussmaul, 1995). The first heavily penalises basic errors resulting from the confusion of lexical items, ignorance of grammatical rules, lack of basic vocabulary, etc. The second focusses on the *communicative function* of the word, phrase or sentence in question, and makes intelligent assessments of any lexical items that are best omitted or any essential information that needs to be added to the target version. Both models assume that the translator *understands* the source text itself. When this is *not* the case, their relevance is not so easily established. Nor are the two methods of translation quality assessment so easily differentiated.

The disk unit *Lexicography 7* selects examples where a different look at translation proficiency is necessary, one that centres on the meaningful selection of terminology. It takes account of the fact that technical translation into English is carried out in many countries by *non-native* English speakers, and analyses typical mistakes made by German speakers alongside those of non-technically minded, native English speakers. Examination of the disk glossaries reveals the most difficult *fundamental terminology* translators have to cope with, especially the large number of *polysemes* present in science and engineering. The basis chosen for comparisons of fundamental technical translation proficiency and the initial assessment of translation difficulty, in the unit concerned, is therefore the book itself.

Normally in the business world it is cheaper in the long run to employ experts to resolve one-off technical problems rather than try to make do with the facilities at hand. But sometimes faulty translations by in-house employees with a limited knowledge of the foreign language and no linguistic education whatever are preferable to those of well educated external translation experts simply because they appear to make more sense — at least at *phrase level*. The disk unit restricts analysis of translation quality/difficulty to this level. The section below summarises the techniques proposed.

19.2 Phrase-Level Assessment

The disk section analyses three types of typical translation error and proposes a grading scheme involving penalty points. The first type discusses errors made by truly abysmal technical translators involving translations of three common fundamental German expressions from the fields of electrical, chemical and computer engineering: *Binärzeichen*, *Verbindung*, *elektrischer Spannungsimpuls*. The second type involves the more experienced translator who makes a reasonable guess, based on a valued judgement, as to the English equivalent of *Spulenwiderstand* in a particular context. The guess may be wrong because the expression selected does not correspond to the standard term employed by technologists in the field concerned, but the customer is more likely to be satisfied with work done by this person than with translators who persistently make mistakes of the first type.

The third error type requires a more subtle technique of analysis as the margin of error depends directly upon the translator's choice of sentence structure. The example concerns translations of the German semiconductor materials expression *Leitfähigkeit* in the sentence:

> *Ein Kristall, bei dem die Leitfähigkeit auf beweglichen Elektronen beruht, nennt man einen n-Halbleiter.*

There are several alternative ways of rendering *Leitfähigkeit* in the sentence frame below:

> The term *n-type semiconductor* denotes a crystal in which *(the) conduction/(the) conductance/(the) conductivity* depends on mobile electrons.

all of which could be regarded by the translation customer as sensible and correct. But a slight amendment to the sentence formulation reduces the possibilities to two:

> … a crystal whose *conductance/conductivity* depends on …

The substitution *conduction* is not possible here, nor are considerations of relevance relating to the article *the*. A further minimal amendment reduces the possible substitutions to just one:

> … a crystalline material whose *conductivity* depends on …

In isolation, it is not clear whether the German sentence refers to a *process*, a *property* or a *parameter* associated with the concept *Leitfähigkeit*. The first translation is equally vague, the second restricts the interpretation to just *property* or *parameter*, and the third to just the *parameter*.

In a student examination, great care is necessary to check whether the candidate's renderings of individual sentences, which may differ from those of the translation assessor, nevertheless form a consistent interpretation as a whole. The grading of student translators of the third type, some with a talent for remaining vague, others with a more confident decisive approach, would employ Kussmaul's *professional translator view*, where penalty points should differ radically from the penalties for totally ridiculous renderings of *Leitfähigkeit* such as *"*power to conduct*", *"*conducting ability*", *"*conduction capacity*" copied blindly from the first dictionary at hand.

Some university teachers concentrate on style and expression and virtually ignore terminological mistakes when the terminology "was not covered in class". In such cases, the student's diploma provides no preparation for the real world. It merely reflects his or her ability to pass an examination. Other teachers deduct one point regardless of whether the expression chosen is a less preferable alternative or whether it destroys the *message* of the source text completely. This can also produce misleading examination results giving no true reflection of translation ability. The disk suggests more accurate methods of assessing translator competence from the viewpoint of the potential employer or customer, via a balanced scheme that provides accurately weighted results regardless of the degree of difficulty of the source text.

19.3 Other Assessment Criteria

The examples of the previous section illustrate errors made by student or other translators that result from dictionary substitutions and *seriously* impair the quality of a translation. For less serious mistakes, those which may slightly irritate or mislead the translation customer but not lead to a total breakdown in understanding, other criteria of error assessment are employed. Rather than go into depths at this stage, however, examples are discussed that highlight additional considerations necessary for the accurate assessment of translation proficiency among individuals who have moved beyond the stage of simple foreign-language substitution errors.

The following extract appeared in a German student examination paper and was translated as follows:

> *Öffnet man eine Flasche Sprudel, so entweicht ein Gas, das Kohlendioxid (CO_2). Es wird in der Umgangssprache oftmals als "Kohlensäure" bezeichnet.*
>
> Student A — *When a bottle of mineral water is opened a gas escapes, carbon dioxide. It is often referred to colloquially as carbonic acid.*
>
> Student B — ... *In the German colloquial language it is often designated carbonic acid.*

The first student has a good command of English but has overlooked one essential aspect of translation, communicating the message of the source text. The expression *Kohlensäure* involves what English speakers might loosely refer to as *gas bubbles* or *fizz* and is certainly not translatable as *carbonic acid.* Student B does not have such a fluent command of English but has at least realised the problem. If the sentence were correctly formulated, for instance

> ... *In German-speaking countries it is often referred to as "Kohlensäure", a chemical expression implying carbonic acid.*

it could be argued that Student B is a better translator.

Student C left out the second sentence completely, which oddly enough is an even better solution as it was clearly of no importance to the potential translation customer how *German* speakers describe this phenomenon. A fourth student came up with a more ingenious solution —

> Student D — *When a bottle of mineral water is opened it becomes fizzy and a gas bubbles to the surface, carbon dioxide.*

In conjunction with a grading system of points deducted for serious terminological errors, student A should certainly lose a point for this obvious blunder, whereas student B's language error might be cancelled by his or her *perception* of the translation difficulty. Students C and D should receive one or more *plus points* that help to balance out other deficiencies elsewhere in their translations.

This example illustrates a simple translation problem, but one which frequently occurs in a variety of manifestations. Consider another sentence from the same German source text:

> *Man bezeichnet die Lösung als Salzsäure, da Chlorwasserstoff früher ausschließlich aus Kochsalz hergestellt wurde.*

The students translated the first part of the sentence correctly and offer different solutions to the second part:

> Student A — *The solution is termed hydrochloric acid because it used to be manufactured from cooking salt.*
>
> Student B — … *It used to be manufactured from cooking salt.*
>
> Student C — … (second part omittted)
>
> Student D — … *It used to be manufactured from common salt, sodium chloride.*

Student A again fails to notice any problem. Though the translation might look like perfect English at first sight, it does not really make any sense at all. Student B realises the problem, omits the apparently offending word *because* and makes two sentences. Student C omits the second part of the sentence completely but this time loses a small component of the message. Student D follows B's approach but makes a small improvement, an attempt to link *sodium chloride* with *hydrochloric acid* in a similar way to that of *Salz/Salzsäure*.

In a suggested marking scheme, student A should lose two or more points for a serious blunder, whereas C perhaps loses only one, for the small information deficit which may have been important for the translation customer.

Student B remains level. Student D is the only winner with a plus point for ingenuity.

The four students illustrate four distinct types. Student A is an excellent linguist but a poor translator, B a less proficient linguist but potentially a better translator (his or her English might improve). Student C has the motto — *if in doubt leave it out.* This person is likely to be a successful professional translator, as little time is wasted worrying about incidentals, but needs to check the end result of each assignment quite closely. Student D is definitely the best linguist and the best translator and should receive the appropriate academic credit. But unless D is a true genius, speed could work against ingenuity. If D dithers too much on incidentals, C may be the more successful translator in the real world.

Some university teachers feel that a distinctive line should be drawn between *language teaching* and *translator training*, but in practice this rarely happens. The disk is necessary for *technical translator training* and most of it concentrates on *technical language teaching* as this is essential to native as well as non-native English-speakers. It could be argued that the technical collocation dictionary (TCD), which deals *with technical verbs/adverbs/adjectives* and *general vocabulary* used in *technical contexts*, should employ more examples illustrating general translation difficulties like those above. It could also provide longer examples and specify different *text types*, *customers*, etc., with explicit distinctions between essential and non-essential meaning or contrasts given between draft, high-speed and polished translations. Another improvement would be to provide examples of *all* technical terminology used in context, instead of just selected polysemes. But these improvements, though useful, would inevitably entail another project and a second disk.

Unit 20

Concept Analysis

The different features of the organisational structures presented in the two disk volumes can be summarised broadly as follows:

i. the TPD focusses the reader's attention on *polysemy* or *homonymy* and the potential translation problems resulting;
ii. the Thesaurus clarifies *denotations* and *connotations* of troublesome polysemes by relating them to other terms in the same glossary;
iii. the TCD illustrates *usage* and reveals *syntagmatic relationships* among terminology and general technical vocabulary;
iv. the hierarchic term lists focus on *knowledge structures* associated with the microterminologies of individual engineering areas.

No single concise dictionary structure can fulfil all these objectives simultaneously, but *hierarchic organisation*, whether implicit or explicit, is a necessary keystone of any good technical dictionary.

This unit assesses the merits of various organisational structures, individually and collectively, as translation tools. It illustrates intrinsic limitations common to all dictionary structures and proceeds to a general discussion of the problems of concept differentiation.

20.1 Hierarchic Organisation, Compactness

Translators using dictionaries like to obtain the correct target-language expressions for unfamiliar terms rapidly. Lexicographers therefore try to present as much conceptual and contextual information as possible in the minimum space. The glossaries of the book take this objective to different extremes.

The TPD provides rapid access to target-language terminology and occupies roughly half the space of the Thesaurus (one line per entry on a printed page, as opposed to two), but the conceptual information provided is less specific. Access to terminology in the Thesaurus is still rapid, and it is *conceptual* information itself which helps the translator locate correct target-language equivalents in cases involving polysemous source language terminology, rather than didactically inspired *morphological* arrangements as in the TPD. The process of obtaining *micro-terminologies* for specific areas like *circuit components* or *carburettor systems*, however, requires a bit more effort on the part of the dictionary user than looking at hierarchic glossaries. The *hierarchic term-list* approach reduces the physical space of the thesaurus back to one line per entry, and by providing ready information on semantically associated terminology, it narrows the source concept almost as specifically as a lengthy written definition. This approach even provides "slots" for concepts that exist in one language but not in another, the phenomenon of lexical gaps discussed in earlier units.

The engineering chapters employ hierarchic lists to supplement the conceptual information given. The dictionaries of Volume 2 have hierarchic aspects too. The Thesaurus is an alphabetic listing of a multidimensional polyhierarchic terminological structure covering the key concepts of the engineering disciplines. The TPD overrides conceptual boundaries by grouping morphologically similar but semantically unrelated terminology together as entries, but where the entries contain large numbers of sub-entries (related compound terms) the dictionary employs hierarchic organisation in the form of secondary indentation. In addition to this, both the TPD and the TCD employ thesaurus descriptors to indicate hierarchic associations, for particularly troublesome entry concepts. Each of the dictionaries tries to concentrate the maximum information within the smallest physical space. By providing cross-links to the other dictionaries they avoid repetition of identical conceptual information, reduce sets of entries to a manageable minimum and provide a consistent, didactically organised, terminological data-base for fundamental areas of science and technology.

All dictionaries have their limitations. The main problems encountered by technical translators equipped only with conventional alphabetic glossaries of scientific and engineering terminology are overcome by the inter-related, structural lexicological approaches adopted in the book. The rest of this unit discusses the limitations that remain.

20.2 Conceptual Incompatibilities

Occasionally in technical translation, a situation is encountered where slightly different conceptual structures in the languages concerned result in different terminological structures. For instance, the hierarchy below:

1	(railway vehicle)	*Eisenbahnwagen*
11	truck, waggon	*Güterwagen*
12	coach, carriage	*(Wagen)*

indicates to a native British-English speaker that the German term *Eisenbahnwagen* covers not just *trucks*, which are used for transporting *goods* or *freight*, but also *railway coaches* that are for *passenger conveyance*. The superordinate concept *"railway vehicle"* is an artificial one for the British translator, who would normally consider the concept *train* as denoting *"engine + trucks"* or *"engine + coaches"* as the case may be, without labelling the unspecified case. Hence * *"engine and vehicles"* is not a normal expression in the British technical language of railway trains. Likewise, German-speakers feel no urgent need to distinguish the concept *coach* from *truck* in normal terminology. Terms like **Passagierwagen*, **Personenwagen* are artificial constructs.

German labels the superordinate concept in the example above, whereas English gives concise labels to the subordinate ones. The terminology structure of American English is different again:

1	railroad car	*Eisenbahnwagen*
11	freight car, freight wagon	*Güterwagen*
12	passenger car	*Wagen*

The situation is complicated by the fact that German sometimes uses the expression *Wagen* and sometimes *Waggon*. These expressions do not necessarily relate directly either to the American terms *wagon*, *car* or to the British terms *truck*, *carriage*:

Güterwaggon	goods truck (Br.), freight car (Am.)
Schlafwagen	sleeper, wagon-lit, sleeping car (Am.)
Wagen 1.Klasse	first-class coach

Furthermore, after analysing what seem to be the respective connotations of *wag(g)on*, *truck*, *coach*, *car*, *carriage* the translator soon discovers exceptions:

Gepäckwagen	luggage van (Br.), baggage car (Am.)
Güterwagen	goods van (Br., u: passenger train)
Speisewagen	dining car (Br./Am.)

There is a final irritation for terminologists. The term *sleeper* is homonymous. The same expression refers to part of the railway line (Ge. *Eisenbahnschwelle*).

In cases like the above, even the most accurate conceptual dictionary structure breaks down. The lack of morphological consistency is too great and multidimensional representation of hierarchies becomes too difficult on the printed page. Concepts are intertwined. Fortunately most cases of conceptual incompatibility arising in technical language (*Spannung*, *Widerstand*, etc.) are easier to resolve.

20.3 Contextual Equivalence

The fact that the technologies of railway systems grew independently over long periods in Germany, Britain and the United States, with little interaction between engineers in the various countries, leads to terminological inconsistencies and conceptual incompatibilities. Similar problems occur in texts dealing with household electrical, heating or plumbing systems. In such cases, the same German source text may require a different translation according to whether the product concerned is to be marketed in Britain or North America. Chapter 8 reveals a third difficult area: Automobile Engineering.

Problems are of two distinct types. The first is easily resolved. Different expressions for the same concept, for example:

	Br.	*Am.*	*Ge.*
1	generator	generator	Lichtmaschine, Generator
11	dynamo	d.c. generator	Gleichstromlichtmaschine
12	alternator	a.c. generator	Drehstromlichtmaschine

merely leads to slightly bulkier dictionaries. In the Thesaurus, it leads to a few double entries. The same applies to the TPD. The second problem, namely where a labelled concept in one language has no label in the other, is more difficult for translators to cope with. For instance, the electrical terms (Chapter 2):

1	(impedance)	Widerstand
11	resistance	Verlustwiderstand
12	reactance	Blindwiderstand
13	impedance	Scheinwiderstand

The English-speaking technologist regards *resistance* and *reactance* as special limiting cases of the concept *impedance*. The term *impedance* is defined as a combination of *resistance* and *reactance*. German defines the superordinate concept *Widerstand* and specifies three distinct types. The reluctance of German electrical engineers to adopt the recommended, internationally understood expressions *Resistanz, Reaktanz, Impedanz* results partly from their different way of visualising the concepts concerned. Translators who study engineering literature only in one language are not equipped to recognise such problems, and translation errors involving expressions like *Widerstand* are virtually inevitable.

Close examination of the above example reveals that, in the strict sense, it is a mistake for lexicographers even to assume that the German concept *Widerstand* is interpreted by three different English equivalents. *Resistance, reactance, impedance* are not semantically but *contextually* equivalent to *Widerstand.* An English equivalent to the true, superordinate, technical German concept does not exist.

20.4 Concept Differentiation

The hierarchic term-list approach to glossary editing is more likely to guarantee *systematic coverage* of subfield terminologies than other dictionary structures but it is impracticable on a large scale. Apart from access difficulties, there are the problems of incompatibility in knowledge structures *between* languages and the multidimensionality of concept structures *within* language to be considered. The thesaurus and polyseme dictionary approaches overcome some of these obstacles but face the usual difficulties of terminology coverage. Conventional alphabetic dictionaries provide the worst of both worlds and a poor dictionary often reflects the lexicographers inability to think hierarchically when differentiating closely related concepts. But the ultimate hinderance to any dictionary structure seems to stem from the limitations of conceptual organisation itself, a problem dramatically illustrated above in the example:

Eisenbahnwagen, -waggon car, carriage, coach, truck, van, wag(g)on

This section takes a close look at hierarchic organisation and the resulting lexicological implications.

Consider another example of purely contextual (i.e. non-semantic) equivalence, the following extract from the Polyseme Dictionary:

Zange *f*:

Abisolierzange	TOOL	cable strippers

Drahtzange	TOOL	wire cutters
Flachzange	TOOL	flat-nose pliers
Kneifzange	TOOL	pincers
Universal-Gripzange	TOOL	mole wrench

The morphological properties of German enable these terms to occur at the same entry in the TPD, and at first sight they seem to be hierarchically related to a common superordinate concept "*Zange*". But this simple interpretation is an illusion. First, the superordinate concept has no equivalent term (i.e. no *label*) in English. *Zange* in isolation has no translation of its own and can only be expressed by subordinate terms:

Zange cutters, pincers, pinchers, pliers, tongs, strippers, etc.

Secondly, even before major differences in language structure are considered, the translator needs to realise that though *pincers*, *pinchers*, *pliers*, *tongs* occur in isolation, *cutters*, *strippers* normally occur only in compounds: *cable cutters*, *barbed-wire cutters*, *steel-mesh cutters*, *insulation strippers*. Thirdly, the concept *wrench* does not seem to belong here at all. An English-speaking technologist sees little or no correlation between *wrench* and concepts like *pincers*, *pliers*, *wire cutters*. Grammatical considerations bear this out too. *Wrench* is an ordinary countable noun (CN), whereas the other terms are pair nouns (PN). They have no singular form and co-occur with the expression *pair*: "*a pair of pliers*", etc.

Degrees of difference in the knowledge structures of English and German reflect the seriousness of translation errors. An electronics enthusiast reading a translation containing the incorrect expression **wire tongs* would probably smile to himself, make a mental substitution **wire grippers* and immediately infer the correct concept *pliers*. On the other hand, on seeing the expression **grip tongs* (instead of *mole wrench*) a household plumber, automobile mechanic or any other engineer would remain baffled.

Technologists faced with an arbitrary concoction of translation errors involving expressions like *resistance*, *reactance*, *impedance* interspersed with *resistor*, *resistivity*, *reluctance*, all of which could co-occur within short stretches of an electrical engineering text, are equally baffled. The TPD highlights this type of problem. The Thesaurus and the engineering chapters attempt to deal with it. But they cannot cure it. In cases of doubt and especially for problems not covered by the book, the only truly safe approach for the translator is to acquire relevant literature in German and in both British and American English. Good translation detectives examine all available evidence first-hand.

20.5 Concept Recognition

On encountering so many technical polysemes in German the reader may get the impression that the language is one big mess and draws upon a much smaller vocabulary than English. But things should be kept in perspective. Areas like Steam Turbine Engineering and Satellite Communication are so remote from each other that engineers barely realise that similar expressions are used for entirely different concepts (*Impuls*: momentum, pulse; *Spannung*: stress, voltage; *Leistung*: power, performance). Moreover, Germans are generally aware of international trends and happily incorporate anglicised or other expressions into their language to avoid confusion. But "avoiding confusion" refers to native-speaker technologists themselves, not to translators. An engineering author or report writer who feels that the expression *Widerstand* in a particular sentence may be misinterpreted by his colleagues might qualify this to *Scheinwiderstand* or *Impedanz*, but will not bother otherwise. Such substitutions can confuse translators even more when they find the terms *Impedanz* and *Widerstand* used synonymously in one part of a report but differently elsewhere in the same report. These problems occur in general translation too. It is not the language that is responsible. They are simply more difficult to cope with in engineering situations.

The lexicography units of the e-book deal with problem-solving strategies based on the application of *general linguistic* principles. This unit takes the reader into *general semantics*. Translators who, in the past, would not have bothered to even consult a dictionary for terms like *Impuls*, *Leistung*, *Widerstand* now see the concepts in a different light. The objective of the unit, however, is not just to extend the reader's interest for semantics itself, but to improve his or her translation abilities in regard to fundamental technical terminology. The next section resolves one final example of translation difficulties that occur when knowledge structures differ radically, an example encountered frequently throughout the book, the various interpretations of the German concept *mechanische Spannung* (compression, compressive stress, compressive force, stress, tensile force, tensile stress, tension).

20.6 Concept Splitting

Construction engineers use the terms *tension/compression* in complex calculations and computer models of the *load patterns* relating to structures like scaffolds, bridges, roof support systems, greenhouses. The structures are composed of individual *members* (Ge. *Bauglied*), where the terms *tension/compression* denote the *tensile/compressive forces* present in the members under particular loading conditions. These expressions relate to the stability of a given structure (Chapters 1, 13). The same engineers use the terms *stress, tensile stress, compressive stress* to denote the effects of particular *loads* or *loading configurations* (Ge. *Belastungen*) on the *materials* used in the structure. *Stress* produces *strain* which results in elongation, compression, bending or twisting of the members. The terms *stress, strain, tension, compression* also denote *physical quantities* used as parameters for measuring *stress, strain,* etc.

An elaborate hierarchic arrangement could be proposed for the above concepts and their respective parameters, but it would not help the translator, as German uses the term *Spannung* for just about everything. Instead, it is helpful to know that the parameters *stress* and *tension* correspond to different physical quantities and therefore have different units: $newton.m^{-2}$ as opposed to *newton.* Close attention to context may resolve whether the stresses implied are *tensile* or *compressive.* This is the method proposed in Chapter 1.

Another method is introduced here. It employs hierarchical thinking and draws parallels with other examples in this unit: *Eisenbahnwagen, Widerstand, Zange.* Consider the following lexical arrangements:

1	(...)	1	stress	1	Spannung
11	tension	12	tensile stress	12	(Spannung)
12	compression	13	compressive stress	13	Druckspannung

Technical English employs *tension* and *compression* as diametric opposites of the type *positive/negative charge.* No term covers the general case as there is no concept for this. German, on the other hand, designates the general case *Spannung* and one specific type *Druckspannung.* No term covers the diametric opposite. German-speaking engineers substitute *Spannung* to imply *tension* and qualify this to *Zugspannung, Zugkraft, Spannkraft* only in specific cases when they themselves anticipate confusion.

Tension and *compression* are *forces.* Where these concepts occur with a plural meaning the terms *tensile/compressive forces* appear as alternatives. German employs alternatives too — *Spannungskräfte, Druckspannungskräfte,* but only in very specific circumstances. The normal expressions are again *Spannung/Druckspannung.* As if adding to the translator's nightmare, German employs the same terms for a different concept with different units: *stress.*

Unlike *tension*, the term *stress* relates to the *superordinate* concept and, in this case, it is directly equivalent to its German counterpart *Spannung*.

Familiarity with the subject field is a tremendous advantage here. But, in such cases, translators are well-advised to discuss the source text with the customer first. Good translation is impossible unless the intentions of the author and his visual conceptions of the realities concerned are properly understood. If the customer is aware that the translator does not have a hope in hell of rendering terms like *Spannung* correctly, a brief discussion or a few words of guidance can vastly improve matters, provided that the translator knows the right questions to ask.

Unit 21

Mathematics

Beginning with fundamental *physics* conceptions, the book gradually develops the reader's conceptual understanding and command of technical terminology to a point where it covers the underlying aspects of all the main branches of *science* and *engineering*. By comparing and contrasting subject fields, it develops the reader's awareness of the polysemy of technical language. The final disk chapter, Chapter 16, adopts a similar approach but concerns not an area of technology itself, but one whose terminology is persistently borrowed by all scientific and engineering fields, namely *Mathematics*.

The sections below take readers back to their early school days and try to rekindle mathematical understanding. The chapter itself progresses to more advanced areas like *probability distributions*, *vectors* and *infinitessimal calculus*, and provides details of where and how these fields relate to engineering. Useful microterminologies of fundamental mathematical concepts like *circle*, *graph*, *number*, *polygon* appear in both languages in Figures 16A–G. For elementary fields, such as *geometry*, these serve simultaneously as a basis for the discussion of hierarchic systems of terminological organisation in general.

Mathematics is a difficult area for readers to absorb merely by peering at a monitor screen. The initial sections of the disk chapter are therefore reproduced in this unit.

21.1 Fraction, Proportion, Quotient, Ratio

The German mathematical expressions *Anteil*, *Bruch*, *Quotient*, *Verhältnis* are frequently mistranslated. Examples taken from elementary mathematics textbooks, such as:

Der Anteil der Schüler in Klasse 1b ist größer.
Schreibe als Bruch (kürze wenn möglich): 16% 35% 80%.

Berechne die Quotienten U:I.
Das Verhältnis zweier Zahlen entspricht einem Bruch.

reveal the basic correspondences *Anteil* (proportion), *Bruch* (fraction), *Quotient* (ratio), *Verhältnis* (ratio). But other examples soon reveal exceptions:

Das Vergleichen von Anteilen wird erleichtert, wenn man die Zahlenangaben auf eine gemeinsame Grundzahl bezieht.
Fractions are easier to compare when the numbers involved are referred to a common denominator.

Strom und Spannung nehmen im gleichen Verhältnis zu.
The current and voltage increase proportionally.

The semantic connotations of the English technical term *ratio* overlap with those of the German expressions *Quotient, Verhältnis,* but only partially. Similarly, *proportion* overlaps partially with *Anteil, Verhältnis, Proportion.* And there is an English term *quotient,* which does not have the same connotations as the German *Quotient.* To avoid errors, the translator must pay close attention to *context* and observe syntactic distinctions governing the *usage* of this terminology.

English employs expressions like: a *fraction* of 5/9 ("five ninths"); a *proportion* of 45 out of 81; a *ratio* of 5 to 9 (written "5:9"). It is interesting to note that these three expressions are in fact *mathematically equivalent.* This does not necessarily mean that they are translational equivalents, however. The translator's decision would depend on the degree of mathematical sophistication of the given source text, on the customer or reader intended, and above all on the particular meaning implied.

The above examples, which involve target expressions revolving around the German terms *Anteil, Proportion, Verhältnis,* indicate certain *syntactic rules* imposed by translation:

fraction *of …*
proportion *of … out of …*
ratio *of … to …*

and a small semantic distinction that may not be immediately apparent to translators:

a ratio of 5 to 9
the ratio of voltage to current

The first expression involves *numbers* and the answer itself is a number; the second involves *variables*, in this case *physical quantities* measured in specific units, and the answer is itself a variable. This distinction can differentiate the German expressions: *Quotient*, *Verhältnis*. Another point of confusion among translators is the fact that, in science and technology, ratios of *variables* (e.g. V:I) are often written like *fractions* (i.e. V/I) but nonetheless refered to as *ratios*. Dictionaries containing entries like:

Anteil	fraction, proportion
Quotient	quotient, ratio
Verhältnis	proportion, ratio

are therefore not necessarily wrong. The information provided is simply misleading.

Certain comments are appropriate, at this stage, on the mathematics terms *fraction* and *quotient*. *Fraction* can imply *proper fraction* (Ge. *Bruch*), two numbers which are divided (e.g. 17/22). It can also imply *decimal fraction* (Ge. *Bruchzahl*, e.g. 0.451). There is an English term *quotient* with no exact German equivalent and a very restricted mathematical meaning implying the result of dividing two *whole numbers*, leaving a possible *remainder*. For instance, when 29 (the so-called *dividend*) is divided by 3 (the *divisor*) the result is the *quotient* 9 with *remainder* 2. This narrow, purely mathematical significance does not justify the widespread inclusion of *quotient* in normal specialised dictionaries for scientific and technical use.

As a summary of this section, translators might care to bear in mind the following general guidelines:

i. *fraction* is not used in association with *variables* (statements like *"the fraction voltage/current" violate grammar rules of technical English);
ii. *proportion* is used in connection with *percentages*, *decimal fractions* and *unspecified amounts* (e.g. "a proportion of the free electrons");
iii. *ratio* has a broader significance in technology than in mathematics and appears mainly in connection with *variables*;
iv. *quotient* is hardly used in science and technology at all.

21.2 Term, Variable, Expression, Function

Consider a simple formula such as $3x^4 + yz$. A mathematician, scientist or technologist regards this as an *algebraic expression* (Ge. *Term*) involving two *terms* (Ge. *Summand*) separated by a *plus sign*. Similarly $1+x+x^2+x^3+\ldots$ is a *series expression* involving an infinite number of *terms* (Ge. *Glied*). German mathematicians employ the word *Term* to mean not *term* but the mathematical concept *expression*. Evidently, English mathematicians borrow a word directly from the general everyday language whereas German adopts a false friend, possibly from early French mathematics (Fr. *terme*).

Another simple series is $1+2x+3x^3+4x^4+\ldots$, which contains the *coefficients* 1, 2, 3, 4, etc. and involves the *variable* x. Taking a practical example, the equation $s=v_0.t+(1/2)gt^2$ concerns the vertical distance *s* fallen by an object or missile dropped or fired from a tower; it involves the variables v_0 and t, the *initial velocity* and the *time elapsed*. The coefficient of the term $+(1/2)gt^2$ contains a *constant* g which denotes the acceleration due to gravity. Here German employs similar terminology: *Variable, Koeffizient, Konstante.*

The words *term, variable, coefficient, expression, sign* (Ge. *Vorzeichen*) reappear throughout mathematics, science and engineering. They are mainly used in connection with *formulas* and *equations*, but also with so-called *inequalities* (Ge. *Ungleichungen*) involving symbols like >, <, ≥ ("greater than", "less than", "greater than or equal to"). The word *formula* can be a contextual synonym of *equation*, in which case it implies an algebraic result derived after application of complex mathematics or profound scientific reasoning. Thus $E=mc^2$ (Einstein's Equation), $R=V/I$ (Ohm's Law) are formulas used in engineering. Or it can imply part of an equation, for instance mc^2, V/I.

Algebraic expressions denote *functions*, which can be illustrated in the form of a *graph*. Thus $v_0 + gt$ denotes the function $v(t)$, the velocity of a free-falling object with respect to time. Functions can also be *series*. The function denoting a sinusoidal curve, representing for instance a *sound wave, mechanical vibration* or *a.c. voltage waveform*, is expressed by mathematicians as $sin(x)$, a function equivalent to the infinite but convergent series:

$$x-x^3/3!+x^5/5!-x^7/7!\ldots$$

where the symbol 5! (etc) represents the value "*factorial* five", the number calculated by 5×4×3×2×1, i.e. 120.

Figure 16A presents a hierarchic list illustrating the significances of the concepts discussed above. The conceptual differences between the coincidental homonyms *term* in English and *Term* in German are apparent here. Inter-relationships existing among the mathematical conceptions *expression, function, variable, coefficient* are also indicated, and so are the semantic relationships (partitive relationships) *term/expression, term/series* (Ge. *Summand/Term, Glied/Reihe*).

21.3 Average, Mean, Variation, Deviation

Functions appear throughout science and engineering in many different forms. But exact mathematical answers to technical problems are not always possible. In areas such as Materials Science or Nucleonics, where predictions are required of the behaviour of tiny particles (*electron, photon, baryon*, etc.), scientists can only state respective *locations, velocities* or *energies* in connection with *probabilities* conforming to the particular *event* or *outcome*. Hence, technical documents tend to contain not precisely measured values of physical quantities such as *energy* but graphs of energy against probability, so-called *probability distributions*. Probabilities can be expressed as percentages but are normally given values lying between one and zero. Thus an event with a probability of 0.001 is extremely unlikely, one with probability 0 is completely impossible and, at the other extreme, one with probability 1.0 is certain and definite. In documents concerning *probability distributions*, certain terms are confused by translators: *average/mean, variation/deviation, relative frequency/probability*, etc. The chapter draws attention to areas of confusion.

Probability Theory and the related field Statistics try to predict the *relative frequencies* (Ge. *relative Häufigkeit*) of particular events by analysis of the *set of outcomes* (Ge. *Ergebnismenge*). The *mean* (Ge. *Mittelwert*) of a set of possible *velocities, energy levels*, etc. is a kind of average value and represents the value "expected" of the variable concerned, the *expectation* (Ge. *Erwartungswert*). The *mean* can be calculated in different ways, which leads to expressions like *arithmetic mean, geometric mean, weighted mean, harmonic mean*. The actual velocities of electrons will of course *vary* about the *mean* and the assessment of this *variation* involves a parameter corresponding to the *mean deviation* (Ge. *mittlere Abweichung*). There are different ways of estimating *deviation* and different parameters, the most common of which is the *standard deviation*.

A few simple rules of thumb may assist translators suddenly confronted with this terminology in the middle of an engineering text. If the text involves *probability estimates* rather than *statistical* ones, in other words complex mathematical or scientific predictions as opposed to a summary of laboratory or production-line results, the translation of *Mittel* is likely to be *mean* rather than *average*. Similarly, if the term *Abweichung* obviously denotes some kind of *parameter* the translation will involve *deviation* rather than *variation*. The concepts *relative frequency, probability, mean, deviation* are mathematical *quantities*, with *units* and *dimensions* analogous to the *physical quantities* of science and engineering, parameters like *strain, angle, efficiency* (Chap. 1).

21.4 Geometric Construction

Civil engineers employ complex two-dimensional drawings to determine *tensions, compressions, load distributions,* etc. in *constructional members* (beams, girders, ties, struts). These drawings are akin to the *force diagrams* of Basic Physics. Regrettably for translators, the drawings themselves are also referred to as *constructions.* The term is associated with the method used to determine answers to engineering problems, namely whether the problem is solved *by calculation,* involving (formerly) *logarithmic tables, slide rules, calculators* and *computers,* or whether it is solved *by construction* and involves mainly a *drawing board* and various *drawing instruments.* Nowadays, of course, the engineer's main tool is the computer, that can also function as a *drawing board* (Ge. *Reißbrett*). Consequently, the distinction between the two methods of problem solution is less explicit now.

Technologists employ methods of *geometric construction* not just for *forces* occuring in concrete visible objects, but for any *physical quantities* which behave like *vectors* (Chap. 1). *Navigators* mapping out flight or shipping routes use constructional techniques to determine *velocities.* And there are other, less obvious applications. Electrical and electronic engineers working with *alternating voltages* employ *phasor diagrams* (Chap. 2) analogous to the *force diagrams* of civil engineers. *Voltages, currents* and *impedances* can be determined *by construction* as well as *by calculation,* and the former method is often much quicker. Indeed, any area of technology that employs *vectors* to represent parameters makes use of *constructional methods* to resolve problems or to obtain rapid approximate answers at least. Clarification of the techniques of vector analysis is provided by certain *disk illustrations* in the thumbnail sections *Basic Mechanics, Basic Electrical Engineering.*

Technical literature concerning *geometric constructions* contains in addition to subject-specific terminology (e.g. *voltage, tension, load*) the familiar preliminary-school terminology of Geometry which most readers will have encountered once, at least in their native language — the terminology of *triangles, parallelograms, circles,* and so on. Geometry in the usual sense, as opposed to relating to the solution of vector problems in engineering, is used by *technical draftsmen* and *architects* designing buildings, and by *design engineers* concerned with the final appearance of *any* marketed or manufactured product (bicycles, cars, bridges, etc.).

Figure 16D lists *instruments* commonly employed by draftsmen. Figures 16E–G contain other common geometric terms in both languages.

21.5 Real/Imaginary/Complex Number

Mathematicians like working with *natural numbers.* This is the set of *whole numbers* (1, 2, 3, 4, … 99 … 1024 …, etc.) together with *zero* and the full set of *negative numbers.* For practical problems, however, they are obliged to work with *real numbers.* The figure −8976 is a natural number whereas −8976.005 is a real number. The mathematical rules for combining numbers (division, subtraction, etc.) differ slightly according to the number system. Software designers are well aware of this and *programming languages* generally stipulate different types of *variable* for different systems of numbers. (Here, the term *variable* implies the computing term (Chap. 11) not the mathematical concept). *Natural numbers* in the field of Software are designated *integers.*

Mathematicians also work with what are known as *complex numbers.* These are *real numbers* added together with so-called *imaginary numbers,* real numbers multiplied by a value "*i*" which represents the square-root of −1 ("minus one"). The square root of 16 is 4, of 81 it is 9, and so on. No natural number multiplied by itself can possibly give a negative answer, but introducing the value *i*, defined by the equation $i^2=-1$, indirectly determines a whole new class of numbers, the *imaginary numbers*:

i, 2i, 3i, …, −i, −2i, -3i, …, 0.321i, 234.005i, etc.

Thus *2i* is the number which, when multiplied by itself, gives the result "*minus 4*", i.e. "$(2i)^2=-4$". Note: "$(-2i)^2$ gives the same result. Imaginary numbers can be combined with real numbers to give so-called *complex numbers*:

"1+i", "3+2i", "15–84i", "1.67–0.006i"

The four examples given (which, for clarity, are separated by quotation marks) represent neither stages in a calculation nor in a mathematical derivation. To the mathematician, scientist or engineer, they are *numbers* themselves.

Despite the apparent denotational complexities, this third class of numbers provides a powerful tool for simplifying mathematical calculations. It enables *values* relating to different *dimensions* to be combined and manipulated, as easily as when multiplying or dividing natural numbers. Moreover, complex numbers provide a convenient algebraic method of expressing *coordinates* in a geometric system with, for instance, *x* and *y-axes.* Thus they can also represent *vectors* and other associated mathematical entities, such as *phasors* and *tensors.*

21.6 Vector Models, Alternative Number Systems

Chapter 1 introduces the idea of *vector quantities*, and reminds readers of their probable first encounter with the concept: as 13-year-old children, determining the *resultant* of two forces applied to an object in different directions, such as two strings trying to lift a heavy weight. Using a pencil and paper, vectors were drawn *to scale*, in *magnitude* and *direction*, and combined to produce a *force triangle* or *force polygon* (Ge. *Krafteck*). The resultant was measured using a *ruler* and its angle read off using a *protractor* (Ge. *Winkelmesser*). This technique is applicable to any vector quantities, and was employed for centuries by navigators responsible for guiding ships to their destinations, who applied it to *velocities*, such as: *ship speed*, *current speed*, *wind speed*. Construction engineers, even in pre-Roman times possibly used the same methods too. But for modern engineering purposes, involving larger numbers of vectors and significant variations in their properties, the solution of mathematical problems by *geometric construction* is time-consuming and inaccurate.

Examples in the disk chapters indicate how civil engineers visualise and portray *tension* or *compression* as vectors in construction beams, and how electrical engineers employ similar models to represent *voltage* and *current* as phasors. A small extension of complex number theory enables vectors to be expressed not just in two dimensions, like a geometric diagram, but any number. Three-dimensional vectors called *tensors* are used by materials scientists to analyse *stress patterns* and *strain distributions* in beams and other objects. Cosmologists invoke a fourth dimension, *time*, in calculations relating to gravitational anomalies, black holes, etc. And mathematicians, who can cope with vectors specified in any number of dimensions, have devised other models, not only for engineers but also for economists and business analysts.

Mathematics is a universal *tool* for all engineers, as important to them as a set of *spanners* or *wrenches* to a car mechanic. Technical translation assignments devoid of any mathematical terminology are an exception. Translators should bear in mind that the term *imaginary number*, which derives merely from the fact that -1 has no square-root, is a misnomer. To scientists and engineers, the numbers themselves are by no means "imaginary". Nor do they regard *complex numbers* as even remotely "complex". Applied to vector models, these alternative number systems provide rapid accurate answers to what would otherwise be very complex trigonometric or geometric calculations, and for many design technologies they are indispensible.

21.7 Geometric Configurations

Design engineers work with *circles*, *triangles*, *squares* and other 2-dimensional conceptions, but projects or individual engineering design tasks generally involve *3-dimensional* figures. Even when working with a pencil and paper, technologists persistently think in terms of *geometric configurations* and employ the terminology of their school-day geometry to describe these configurations. Chapter 16 recalls this *terminology*.

Figure 16E reveals that *triangle*, *quadrilateral*, *pentagon* come under the general category of *polygons*, geometrical objects which have *sides* and corners termed *vertices*. A close look at the hierarchy containing the terminology of *triangles* reveals a minor incompatibility in the nomenclature of the two languages. German employs the expression *Schenkel* for each of the two equal sides of an *isosceles triangle* (Ge. *gleichschenkliges Dreieck*). English has no equivalent.

Selecting another example: German makes use of the term *Kathete* denoting one of the sides of a right-angle triangle. The English equivalent *cathetus* is barely used, and would be incomprehensible to most technologists. German mathematicians evidently retain a name for a concept their English colleagues have no further use for. By the same token, English has the neat terms *median* and *centroid* for "the line joining a vertex to the opposite side" and "the point where the medians intersect" (Ge. *Seitenhalbierende*, *Schwerpunkt*) but no term for the intersection point of the *altitudes* (Ge. *Höhenschnittpunkt*). It would seem that certain mathematical rules, e.g. *the square of the hypotenuse equals the sum of the squares of the other two sides* (Pythagorus Theorem), are rendered more neatly in German, but the price for this is that German schoolchildren learning geometry may be obliged to concentrate more on terminology than their English-speaking counterparts.

21.8 Other Areas of Mathematics

The chapter investigates other, more complex geometrical configurations employed by scientists and engineers to visualise mathematical problems, such as *polyhedra*. It outlines terminologies for areas which can suddenly appear in any technical translation project: *trigonometry*, *coordinate geometry*, *matrix multiplication*, and it discusses well-established solution techniques, so-called *calculuses*, for instance *differential calculus*, *integral calculus*, *vector calculus*.

Linguists can venture to this point, though not many will venture beyond.

Some of the older readers may feel embarassed to realise how much elementary mathematics they have forgotten. Yet the concepts covered in the sections above are largely those which any university undergraduate *embarking on* a course of study leading to an engineering career is familiar with. Mathematics really begins at this point. Fortunately for translators the topic rapidly becomes so complex that scientists themselves stop talking to each other in *words* and resort to an internationally recognisable code of *symbols.* The disk chapter deals with the main areas of mathematics applicable in science and engineering, in the hope that translators faced with unfamiliar or obscure terminology in the middle of a technical document will at least have some idea of *where* to look for L2 equivalents.

UNIT 22

Specific Expression

Nouns do not necessarily constitute the main concern of the technical translator when the source text involves expressions from a long-established area, such as Mathematics. *Verbs*, *adjectives*, *adverbs* and even *prepositions* have a tendency to acquire quite specific meanings. The sections below illustrate problems which cannot be resolved by the normal hierarchical approaches of the TPD and Thesaurus to lexicography. Only sensibly organised *collocational* approaches provide the information needed.

22.1 Special Interpretation

One frequent type of translation problem involves general expressions used with specific technical implications. For instance, finding the correct expressions for *aufrunden*, *Dezimal*, *geltend* in the examples below:

Runde das Ergebnis auf 3 Dezimalen./... auf 4 geltende Ziffern.

Despite the immediate appeal of renderings like *"round the result up to 3 decimals", *"... up to 4 valid digits" these translations are in fact meaningless. The equivalent statements in English are as follows:

Round up the result *to* three *decimal places*.
... to 4 *significant figures*.

Similarly, the effort of concentration on complex technical terminology can sometimes result in a careless approach to the rendering of simple words like *wieviel*, *welche*, *weit*:

Wieviel Prozent Säure enthält die Mischung?
Welche Bogenlänge gehört zum Mittelpunktswinkel von 30 Grad?.
In einem gleichschenkligen Dreieck sind die Basiswinkel gleich weit.

Statements like *"how much percentage", *"which arc length", *"equally wide" are completely misleading and quite wrong in the contexts given. The equivalent English statements are as follows:

> *What* percentage of acid does the mixture contain?.
> *What is the length of* the arc *subtending an angle of* 30 degrees at the centre (of the circle).
> The *angles at the base* of an isosceles triangle *are equal.*

Perhaps the most difficult translation problem above, however, involves the verb *gehören.* Technical dictionaries are unlikely to provide the link *gehören* (subtend) because lexicographers assume the German expression to be an item of general vocabulary. And, even if such links are provided they are of no use without carefully selected collocation examples. When such facilities are not available, problems of this type can only be truly resolved by close attention to first-hand mathematical literature in both languages.

22.2 Specialised Verbs

The last section ends with a difficult example. *Subtend* is technical verb and is restricted to *geometry* and similar mathematical fields. *Gehören* is an item of *general vocabulary* used with a special technical implication but, strictly speaking, it is not itself a technical term. German mathematics does make use of technical verbs though, e.g. *erweitern, kürzen, potenzieren, radizieren,* many of which have no direct English equivalent. Consider the statement:

> *Bruchterme kann man erweitern und kürzen, indem man Zähler und Nenner mit demselben Term multipliziert bzw dividiert.*

English has the term *cancel* (Ge. *kürzen*) for the mathematical operation of *dividing* the *numerator* and *denominator* (top and bottom) of a fractional expression by the same amount, but for the alternative operation of *multiplying* top and bottom (Ge. *erweitern*) there is no term. A solution to the problem in this case is to paraphrase and avoid the technical verbs completely:

> *The form of expressions consisting of fractions can be altered by multiplying or dividing the numerator and denominator by the same identical expression.*

In the case of *potenzieren/radizieren,* direct translations do exist:

Potenziere 7 mit 2.	Raise 7 to the power of 2. (answer 49)
Radiziere 64.	Obtain the root of 64. (answer 8)

but they are difficult to manipulate, especially when the verbs are used as nouns (i.e. *das Potenzieren/Radizieren*) or without direct or indirect objects (numbers, variables, etc.). Paraphrasing of related concepts is also difficult. Expressions like *Potenz, Potenzwert* correspond to concepts which English mathematicians clearly understand, but also have no labels for. They are translatable only in context:

die Potenz x^n	x raised to the power n
die Potenzwert von 2^{10}	the value of 2^{10}

Even paraphrasing is not easy. The variable n in the expression x^n is called the *exponent* (Ge. *Exponent*) when contrasted with x, the *base* (Ge. *Basis, Grundzahl*). But n is referred to as the *power* when used with the verb *raise* (above). In isolation it is called the *index* (pl. *indices*).

Problems like the above are not unresolvable. Linguists realise that technical verbs require particular care and that syntactic aspects may differ greatly in the two languages. But translators currently receive little direct assistance from lexicographers.

22.3 Symbol Conversion

Very few translators become truly bilingual in the area of Mathematics. The subject has existed for hundreds of years and nomenclatures have developed independently. Because of this, international agreement on *symbolic systems* for formulating algebraic, trigonometric and other functional expressions and their associated mathematical operations has existed for many years. Teachers of technical translation often insist that students change *symbols* and convert *units* in the final target version. For source texts involving mathematical derivations this is possibly the worst advice they could receive. An intelligent scientist given a poorly translated text can often reconstruct meaning by examining the mathematic processes. If the mathematics is also tampered with, the translation is doomed.

This section completes the reader's initial introduction to the e-book, except for one small area, the final Appendix of Volume 1, the point where the two parallel streams of information begun in the early stages of the book, the *engineering chapters* and *lexicography units*, converge. The disk Appendix looks at general language and touches upon an area of specialised translation *outside engineering*, which is then used as a vehicle to analyse the possible universal applicability of lexicographical approaches adopted elsewhere in the book. The area chosen is Business English.

Unit 23

Non-Technical Specialised Language

Thanks to constant interaction between scientists and technologists on both sides of the Atlantic the differences between British and American engineering terminology are minimal and concern mainly the different spelling conventions. Apart from a few examples from fields like Railway or Automobile Engineering, the handbook barely mentions any differences. This approach to terminology specification is not appropriate for areas of specialised translation outside Engineering, for instance Legal Translation, Economics, Politics, where translations intended for one half of the English-speaking world are sometimes barely comprehensible to the opposite half. Indeed, German universities which offer courses in Legal or Business Translation and provide two options *British* and *American-English* invariably find that students are so confused, they completely ignore one option.

This section takes a small break from Engineering and looks at "technical translation" in the broader sense; to avoid confusion the expression *specialised translation* is reserved for the superordinate concept (Ge. *fachsprachliche Übersetzung*) and *technical translation* denotes the subordinate concept, the discipline relating to science and engineering (Ge. *technische Übersetzung*). The section lists some relatively "non-technical" terminology which non-native English-speakers, and indeed native-speakers unfamiliar with life across the Atlantic, often confuse. It then takes a brief look at the applicability of the thesaurus and collocational approaches to the arrangement of terminologies for other quite different areas of language. Illustrations are provided from general language and from the field of Business Studies.

A more complete version of this unit appears in the *Appendix* of the e-book. The glossaries discussed are located in the following sections:

Figure A1: Transatlantic Lexicon
Figure A2: Transatlantic Thesaurus
Figure A3: Commercial Thesaurus
Figure A4: Commercial Collocation Dictionary

23.1 Language Variants

Even native-speakers only become proficient in language forms other than their vernacular by studying them individually. For the translator, this does not necessarily mean that the German word *See* is rendered as *lake* for an English customer, *loch* for a Scot, and *billabong* for an Australian, but an awareness of the main differences between the two standard variants British and American English is necessary.

A British translator working on an assignment for an American company needs to pay constant attention to the spelling of words like:

colour, labour, odour, vigour (Br.)
color, labor, odor, vigor (Am.)
catalogue, cheque, instalment, guarantee
catalog, check, installment, garantee
dispatch, enquiry, endorsement
despatch, inquiry, indorsement

especially if these terms regularly appear in documents relating to company policy. Spelling discrepancies are mastered by intelligent use of dictionaries. More critical from the translation viewpoint is *terminology.*

Figure A1, the *Transatlantic Lexicon,* lists differences in the *general terminology* of the two types of English. The terms are arranged in subsections beginning with potential engineering areas, such as *buildings, roads, railways,* and moving on to more general areas like *clothing, office* and *household* terminology. A few samples from the lexicon appear below:

Doppelhaus	semi-detached house	duplex house
öffentliche Toilette	public convenience	restroom
öffentliche Schule	state school	public school
Rollo	roller blind	shade
Schulhof	playground	schoolyard
Autobahn	motorway	highway
doppelte Fahrbahn	dual carriageway	divided highway
Kreisverkehr	roundabout	traffic circle
Parkplatz	car park	parking lot
Rastplatz	lay-by	rest area
Sackgasse	cul-de-sac	dead end
Bürgersteig	pavement	sidewalk

U-Bahn	tube, underground	subway
Unterführung	subway	underpass
Verkehrsampel	traffic lights	stop light, traffic light
Zebrastreifen	zebra crossing	crosswalk
Motorrad	motorbike	motorcycle
Mofa, Moped	moped	motorbike
Reisebus	coach	bus
Wohnmobil	motor caravan	mobile home
Wohnwagen	caravan	trailer

The glossary enables German translators to adapt their English according to whether the intended recipient is based in Europe or America, and provides a useful *crash course* in British or American for native English speakers.

The next section illustrates a convenient shorthand technique for noting semantic discrepancies among terminology and forestalling general translation errors resulting from inappropriate mixtures of British and American English.

23.2 Distinctive Feature Specification

The book makes repeated use of a small set of *thesaurus descriptors* to denote *concepts* associated with specific terminology. The same descriptors enable terminology in a dictionary list to be defined in terms of known vocabulary or of other entries in the same list. The subset below:

t:	*a type of*	a:	*associated with*
cs:	*contextual synonym of*	u:	*used in connection with*
ex:	*example of*	ct:	*contrasted with*

is employed in the *Transatlantic Thesaurus* of Figure A2 to distinguish the meanings of certain orthographically identical terms in Figure A1. Again, a small representative sample:

Term	*Implication, Br.Eng.*	*Implication, Am.Eng.*
baton	a: orchestra conductor;	a: riot police;
cabinet	u: lounge furniture;	u: household fittings;
corporation	u: municipal authority;	t: joint-stock company;
cupboard	u: furniture;	u: kitchen furniture;
cuffs	a: shirt;	a: shirt, pants;

motorbike	cs: motorcycle;	t: motorised bicycle;
pants	cs: underpants; u: vest;	ex: slacks, jeans;
pavement	a: pedestrians;	t: road surface;
pitcher	t: large earthenware jug;	ex: cream pitcher;
precinct	t: shopping centre;	t: administrative district;
public school	t: private school;	t: state school;
purse	a: loose change;	t: handbag;
slacks	t: womens' trousers;	t: pants; ex: dress slacks;
stock	u: warehouse; t: goods;	t: investment;
subway	t: pedestrian walkway;	t: underground railway;
suspenders	a: womens' stockings;	a: mens' pants;
tack	u: carpeting, panelling;	u: office stationary;
trailer	u: load being towed;	t: mobile home;
truck	u: rail transport;	u: road transport;
vest	t: underwear;	t: garment;

Concepts are displayed as a neat list of *primary implications* in alphabetic order of the lexeme likely to cause translation difficulties. In contrast to the *Technical Thesaurus*, this glossary makes no attempt to provide more explicit information on the concepts themselves. It assumes that the reader will have come across the terms at some time or other anyway, and employs the thesaurus descriptors either to indicate a *distinctive feature* separating the meanings of identical lexemes or to provide an *example* which helps narrow the concept in one variant or the other.

German-speaking translators will regard this glossary as an unexpected free gift in a book expected to deal only with technical translation. But all readers should pause at this stage, look closely, understand the implications and uses of the descriptors, and memorise them. The same descriptors occur throughout the disk and clarify concepts of a far greater degree of complexity. They appear again in the next subsection, which deals with what, for some readers, may be another unfamiliar area: Business Studies.

23.3 Business Translation

Good specialised translators are aware of the polysemy of natural language, particularly simple adjectives/adverbs, such as *stark* or *leicht*:

starke Verteuerung	sharp price increase
starke Verbesserung	marked improvement
starker Bestelleingang	rush of orders
stark abbauen	draw heavily (on reserves, etc.)
stark zunehmen	increase rapidly/... dynamically
stärker steigen	rise faster/at a faster rate
leichter Überschuß	small surplus
leichte Geldpolitik	easy-money policy
leicht entflammbar	highly inflammable

Some expressions necessitate paraphrasing:

als Vertreter tätig	1. acting as an agent (for a company) 2. representing (a company/department)
in der Stereobranche tätig	1. working in the stereo industry 2. dealing in stereo systems
weltweit tätig	operating all over the world

and some require *specific* (non-synonymous) renderings according to the context:

groß	sizeable, substantial, large-scale
günstig {Preis}	acceptable, competitive, favourable, low
Geschäftsverbindung	business connection, business association, business relationship, business contact

Verbs require appropriate attention too:

Sturmschäden decken	cover storm damage
einen Bedarf decken	meet a demand
einen Fehlbetrag decken	offset a deficit
durch Versicherung decken	back by insurance

and idiomatic expressions may be problematic, especially for non-native speakers:

Produkte führen	supply products
Verhandlungen führen	conduct negotiations
in Kraft treten	become effective
an Kraft gewinnen	gather momentum

Although the examples above have nothing whatever to do with Engineering it is interesting to note that the expressions *stark, leicht, groß* which cause serious problems in technical translation (see TCD) are also problematic in this field. Verbs too, e.g. *decken*, *führen*, *gewinnen* require specific translations in specialised contexts, and terms like *Kraft* which have precise engineering connotations (*force*, Chap. 1) acquire idiomatic renderings like *momentum*. It is interesting to observe other parallels as the discussion continues.

Translators cannot survive without computers and many individuals become keen software enthusiasts. Having established a collocational data base, such translators soon discover interesting cross-collocations. For example, prepositions:

im Durchschnitt	on average
in bestimmten Zeitabständen	at specific intervals
in den nächsten Jahren	over the next few years
in Wert von	to the value of

syntactic contrasts between German and English:

über Verbindungen verfügen	to have contacts (at ones disposal)
auf einem Gebiet spezialisiert sein	to specialise in an area

A collocation dictionary also provides access to frequently repeated phrases requiring a range of translation alternatives:

zum Teil	partly, to some degree, to a limited extent
zu gleichen Teilen	equally, in equal proportions, equal numbers of
ein großer Teil	a substantial proportion, a sizeable amount

and lexical contrasts of moderately specialised word forms appear too:

Branche	line, industry, sector
Haushalt	household, budget
Lieferung	delivery, consignment, shipment
Umsatz	sales, turnover

that help to sharpen the translator's awareness of the different implications of terminology used in the field.

The older generation of translators approached the problem of terminology in specialised language by amassing enormous alphabetic card indexes, the contents of which were eventually published as dictionaries. Concept definitions and subfield specifications were often omitted, either because of the non-systematic compilation procedures employed or simply to save space. But modern data-handling facilities can easily cope with this additional information.

Moreover, when employing thesaurus descriptors, the amount of extra storage space is minimal. For example, the English terminology below:

Kredit	loan, credit
Rabatt	discount, rebate, reduction
Transport	carriage, conveyance, haulage, transit

can be differentiated in a systematic reference system as follows:

carriage (a: cost of conveyance)
credit (u: delayed payment; a: goods received)
discount (t: reduction; u: bulk order, special concession)
haulage (t: conveyance; u: road, rail)
loan (u: money advanced; a: banks, financial institutions)
rebate (t: reduction; u: complaint)
transit (u: period of conveyance; ex: in/during transit)

Figure A3 contains a section from an English-German *Commercial Thesaurus* and Figure A4 shows part of a small *Commercial Collocation Dictionary*, both relating to the broad field of Business Studies. The terminology derives from four main areas: General Economics (e), Banking (b), Commercial Correspondence (c), Financial Reporting (r). Two extracts are reproduced here:

pension provisions {u: balance sheet}	*Pensionsrückstellungen,-m*
promissory note {t: draft}	*Solawechsel,-m*
publicity executive {u: company}	*Werbeleiter,-m*
purchaser {a: order}	*Käufer,-m*
quotation {u: order}	*Preisangebot,-n*
raw material costs	*Materialaufwand,-m*
rebate {t: reduction; a: complaint}	*Rabatt,-m*
receipt {u: goods ordered}	*Empfang,-m*
receivables {u: balance sheet}	*Forderungen,-pl*
receiver {u: order}	*Empfänger,-m*
reduction {cs: rebate, discount}	*Preisnachlaß,-m*
reminder {cs: dun}	*Zahlungserinnerung, Mahnung,-f*
remittance advice {u: payment}	*Zahlungsanzeige,-f*
representative {a: company}	*Vertreter,-m*
reserves {u: assets}	*Rücklagen,-f*
retailer {t: dealer}	*Einzelhändler,-m*

auflösen (b) *Ich bitte Sie, dieses Konto aufzulösen.*
Please close this account and transfer the balance to my savings account.
auflösen (c) *Er löst das Geschäft auf und entläßt alle Angestellten.*
He is closing down the business and dismissing all staff.
auflösen (r) *Im Falle der Auflösung hat man Anspruch auf Rückzahlung.*
In the event of liquidation one is entitled to reimbursement.

aufnehmen (c) *Die Firma möchte eine Hypothek auf ihr Grundstück aufnehmen.*
The company would like to mortgage its real estate.
aufnehmen (c) *Die Geräte werden von unseren Kunden sehr positiv aufgenommen.*
The appliances are getting a warm reception from our customers.

ausgleichen (c) *Wir senden Ihnen einen Scheck zum Ausgleich der Rechnung.*
We are sending you a cheque to cover the invoice amount./… check (Am.)
ausgleichen (e) *Diese Kosten können wir durch höhere Verkaufspreise ausgleichen.*
These costs can be offset by raising sales prices.

befördern (c) *Er ist zum Abteilungsleiter befördert worden.*
He has been promoted to departmental manager./… head of department.
befördern (c) *Man hat ein Anrecht auf Beförderung von 20 Kg Freigepäck.*
One is entitled to a baggage allowance of 20 Kg./… free baggage allowance …
befördern (c) *Unsere Waren werden mit der Bahn befördert.*
Our goods are (being) shipped by rail./… transported/conveyed …

The illustrations reveal that techniques for the training and self-teaching of technical translators demonstrated in the book, namely the determination of *polysemy*, the specification of *concepts* via *thesaurus descriptors*, and the use of *collocation dictionaries*, are equally applicable to the systematic study of other areas of specialised language. They provide for the rapid acquisition of translation skills in any situation.

Many universities train students mainly in general translation, either because suitable staff are not available to cope with specialised material or because it is assumed that once the foreign language has been mastered the prospective translator can master the rest unaided. Other universities offer technical translation courses but with staff only equipped for simple everyday areas such as Automobiles or Computers and who avoid terminological problems by using ready-translated material. In view of the vast spectrum of material, linguists are expected to cope with, especially the enormous range of scientific, engineering, medical, commercial and legal documentation, translators are inevitably forced to teach themselves at some stage. The book provides not just a new approach to the teaching of specialised language, but a new look at the subsidiary work involved in translation activities and sound advice on the systematic handling of everyday lexicological and terminological information.

UNIT 24

Translator Education

The book has one further application. It provides supplementary material for use by translator educators in university courses covering the broad basis of science and engineering translation. The author's own experience over a period of ten years using a book with a similar theme has shown that, contrary to conventional expectations, students can indeed cope with a wide variety of interlocking technical fields, if they are guided by personnel already familiar with the material. This unit therefore proposes some different ways of using the disk as a teaching aid. It relates them to various standard approaches to technical translator training, ranging from the classical *transmissionist* approach of the university lecturer to modern *constructivism*.

24.1 Constructivist Approach

Don Kiraly is one author who makes a radical departure from the classical approach to university translation classes, quote — the *transcoding of sentences amputated from real translation situations*, to one based on extensive inter-student participation. He encourages students to work in groups and discover the techniques of professional translation for themselves, being *guided* in a particular direction by their teacher. With a detailed plan constituting the curriculum of an introductory course in translation studies, Kiraly shows how he has tempered his role as *a sage on the stage* and moved towards being more of *a guide on the side.*

Kiraly begins by describing workshops where students discuss their interpretation of words like *shadowy, drenched, laden* in isolation, in juxtaposition with other words *shadowy peace, sun-drenched, flower-laden*, and at phrase or sentence-level. Another workshop takes examples from a tourist brochure for a German town Bad Dürkheim, discussing interpretations of *Fremde* (stranger, foreigner, outsider) and the connotations of *Bad* (spa, health resort, etc.) in the sentence:

Dürkheim — nur Fremde setzen dem Namen 'Bad' voraus

and whether or not there is any practical justification, in the context concerned, of translating the sentence at all. The examples illustrate certain parallels with this book: the didactic principles governing the TPD; those governing the TCD; certain examples of Unit 19 (*Kohlensäure, Salzsäure*).

The workshops continue and Kiraly reveals how students are encouraged to bring interesting material of their own to be analysed *collectively* according to a scheme of decision categories: *cognitive, cultural, linguistic, textual,* etc. There are many more stages in the translator education process, which ultimately involves guided or supported translation projects with virtually full-fledged independent work on the part of the students. Kiraly mainly focuses on the acquisition of *general translation* skills, but some aspects of his *social constructivist* approach might also be applied to *technical translator* education.

24.2 Social Approach

The disk contains a vast amount of technical information arranged rather like a gigantic teaching programme itself. One way to stimulate digestion of the disk information for future translation assignments is to study the handbook. Another is for the teacher to select L1 material geared to specific chapters and demonstrate features of the disk as a translation tool. A third method is to encourage students themselves to find such material. For example, a student who remembers some school physics may produce an interesting mechanics text from an elementary textbook to be discussed and translated orally by the class. Students immediately realise that Chapter 1 is the place to look. Any holes in their (or their teacher's) understanding become evident as alternative renderings of terms like *Kraft, Leistung, Dehnung* are disputed and checked via the microglossaries, TPD, Thesaurus, illustrations, etc. After the class, students re-read the chapter, re-examine the glossaries and illustrations and see them in a new light, probably retaining much more of the information permanently than they would have done by working alone.

The same school textbook may contain suitable *basic electrical* material for the next class (Chapter 2), or other students may bring along a car manual, a satellite receiver guide or even a child's *chemistry set* instructions for class discussion and translation — with close attention to other chapters and reference sources of the disk. The teacher gradually guides the students through the entire disk. A semester or two later, when a large proportion of the basic material has been covered, the students are ready to progress to more advanced material handled in professional translation assignments. They also know intuitively what essential information relating to these assignments is readily available on the disk, and where to find it immediately.

24.3 Electronic Approach

Kiraly's guidelines for his *social constructivist* approach to translator training encompass the full range of computational facilities. He describes various learning techniques, such as: *one-alone* involving autonomous learning from databases, journals, software libraries; *one-to-one* as with on-line peer tutoring or e-mail tandems; *many-to-many* with discussions, role-play activities, project groups. He contrasts these with *one-to-many*, the so-called *transmissionist approach* to knowledge dissemination, the classical one of the teacher or lecturer at the front of a class.

The e-book can be applied to all these approaches and indeed it should be, coming at the start of a technical translation course before other less easily or less readily available electronic tools are even considered. Kiraly's ideal *networked classroom* with a separate workstation for each student allowing them to call upon their own computer expertise for translation-related research is directly amenable to guidance methods for absorbing information from the disk. Teachers can easily select source material containing disk terminology or involving aspects of the disk information, and students prefer this rapid systematic initial approach to the acquisition of technical translation skills to clicking around aimlessly on the Web.

24.4 Transmissionist Approach

A statement earlier in the book refers to modern linguists changing their area of enthusiasm at regular intervals like flocks of birds. But sometimes they are more like flocks of sheep, when snobbism or fear of being classed among the *cruddy-duddies* causes them to attack new ideas relating to areas no longer in fashion without hesitation or reason. For instance, one statement in response to the disk, made by an early short-sighted critic was "*everything is available on the Web*". That of course is not true. Information of this calibre, carefully structured in two languages for a vast collection of engineering areas, is not available anywhere on the Web, not even in small doses. It seems that while pretending outwardly to favour constructivist guidance methods, the reader automatically interpreted the disk in terms of his *actual* approach to translator training, the classical one, that of a conventional transmissionist.

Oddly enough, the transmissionist approach itself is also adaptable to the usage of the disk. Conventional university teachers who have chosen and prepared material for translation classes, involving disk terminology, during stress-free semester vacations have an advantage over students asked to perform the same task within just a few days. This competitive edge, on which so much of their personal esteem depends, may persist. The teacher should not take this to extremes though. It may be easier to find translation material to suit the disk than choosing material at random and floundering around on the Web for terminology, but translation classes should not entirely neglect other electronic tools. Just as students require a lengthy period to master their foreign languages without the burden of technical considerations, in the same way they require a period of systematic study to concentrate upon *technical language* without the perpetual agitation of Web-searching and the constant flicker of the changing monitor screen.

24.5 Disk Approach

Experience has shown that many university teachers do not like their students to have access to large sources of accurate information, all in one place, because they feel their own integrity threatened. Hence the disk approach to translator education was softened to give teachers a chance to adapt before students discover the true power of the book for themselves. Hasty users, or what this book refers to elsewhere as *non-thinking translators* will regard the disk as a database, an error that at this stage warrants no further discussion. To true linguists, the disk gives the impression of being a didactic tool for promoting *individual* translation expertise, whether the user concerned is a professional translator or someone learning the profession. Indeed, it is just that.

Used sensibly it can also function as either a transmissionist or a constructivist learning tool, one which removes the indecision and bewilderment currently hampering students new to fields of engineering, so that technical translation becomes an intellectual activity as intrinsically rewarding as general translation classes are likely to be. Methods of tuition, tutoring or group guidance vary from one academic institution to another, among individual teachers, and there are many intermediate approaches described by authors on which Kiraly's ideas are based (Nord, Toury, Kussmaul, etc.). But, applied to scientific or engineering translation, they all have one thing in common. Successful transitions from *student* to *graduate* to *professional* translator are smoother and easier by any method using the disk, than any without.

Bibliography

Bibrack, Lautenschläger, Schröter. 1985. *Chemie — Fakten und Gesetze.* Cologne: BZ-Verlag.

Clarke, John S. 1986. *Applied Chemistry.* Sevenoaks, UK: Hodder & Stoughton.

Davies, J. K. 1993. *Space Exploration.* Edinburgh/New York: Chambers Encyclopedic Guides.

Eydam, Erhard. 1992. *Die Technik und ihre sprachliche Darstellung — Grundlagen der Elektrotechnik.* Hildesheim: Georg Olms Verlag.

Felber, Helmut. 1980. *International Standardization of Terminology.* The Hague: Mouton Publishers, Intl. J. Soc. Lang. 23.

Hann, Michael. 1992. *The Key To Technical Translation,* Vols 1 & 2. Amsterdam/Philadelphia: John Benjamins Publishing Company.

Hawking, Stephen. 1997. *A Brief History of Time.* UK/Australia/New Zealand: Transworld Publishers, Bantam Press.

Haynes. *Service and Repair Manuals* (for virtually any motor vehicle or motorbike manufactured since 1970). UK/USA/Sweden: Haynes Publishing.

Kiraly, Donald C. 1995. *Pathways to Translation.* Ohio: Kent State University Press.

Kiraly, Donald C. 2000. *A Social Constructivist Approach to Translator Education.* Manchester: St Jerome Publishing.

Kuchling, Horst. 1985. *Physik — Formeln und Gesetze.* Cologne: BZ-Verlag.

Kussmaul, Paul. 1995. *Training the Translator.* Amsterdam/Philadelphia: John Benjamins Publishing Company.

Lyons, John. 1968. *Introduction to Theoretical Linguistics.* London/New York: Cambridge University Press.

Lyons, John. 1977. *Semantics,* Vols 1 & 2. London/New York: Cambridge University Press.

Murphy, Patrick. 1982. *Modern Mathematics — Made Simple.* London: Heinemann Publishers.

Nida, Eugene A. 1974. *Semantic Structure and Translating.* Munich: Fink-Verlag.

Nida, Eugene A. 1975. *Exploring Semantic Structures.* Munich: Fink-Verlag.

Oates, J. A. 1980. *Lathework — Questions & Answers.* Butterworth, UK: Newnes Technical Books.

Palmer, F. R. 1977. *Semantics — A New Outline.* London/New York/Melbourne: Cambridge University Press.

Reinhardt, Köhler, Neubert. 1992. *Deutsche Fachsprache der Technik.* Hildesheim: Georg Olms Verlag.

Sager, Juan C. 1990. *A Practical Course in Terminology Processing.* Amsterdam/Philadelphia: John Benjamins Publishing Company.

Sager, Juan C. 1994. *Language Engineering and Translation.* Amsterdam/Philadelphia: John Benjamins Publishing Company.

Simon, Hans. 1985. *Mathematik — Formeln und Gesetze.* Cologne: BZ-Verlag.

Wagener, J. Siegried. 1978. *Definition and Origin of the Human Language.* The Hague: Linguistics, Mouton Publishers.

Wüster, Eugen. 1968. *The Machine Tool — An Interlingual Dictionary of Basic Concepts.* London: Technical Press.

Wüster, Eugen. 1974. *Die Umkehrung einer Begriffsbeziehung und ihre Kennzeichnung in Wörterbüchern.* Munich: Nachrichten für Dokumentation.

Appendix 1
Approach Survey

Writing a book which can be read on different levels has great advantages for readers struggling to cope with the vast amount of technical information contained. But for academics it can provide a means of attacking small specific aspects, irrelevant to the main purpose, without really understanding the full book itself. This Appendix puts the record straight by comparing approaches adopted in the disk volumes with ideas of other linguists, lexicologists, semanticists and terminologists of various generations.

1. Convention

Non-native English speakers, may welcome the fact that the simple style employed in the engineering chapters is extended to the linguistics sections with no urgent need for background reading. Other linguists may frown at the author's introduction of new grammatical terms (*pair noun*, *singular noun*), his reluctance to employ the traditional *tree diagrams* to indicate semantic relationships, and his occasional usage of expressions like *attribute*, *element*, *feature*, *field* with implications slightly different from the standard meanings in classical general semantics. But there are good reasons for the approaches adopted.

The objective of these units is to supplement the reader's knowledge of engineering terminology and resolve specific conceptual problems arising in the chapters. The avoidance of tree diagrams saves valuable space, and minimising the number of grammatical categories saves valuable time for readers already struggling to memorise a large set of dictionary labels. The book employs established models borrowed from *general semantics* and may itself make an indirect contribution to the field, but it is intended first and foremost as a translator's handbook, rather than as a contribution to linguistics.

2. Orthography

By an unfortunate quirk of fate, the publication of the book was preceded by a political incident that presented a substantial dilemma. German schoolteachers are obliged by law to adopt a new system of spelling (Ge. *Rechtschreibreform*). The system has been accepted by newspapers, journals and other publications, and twenty-first century Germans will soon write rather differently to their predecessors. The author was faced with the decision of whether to re-edit the entire book and *guess* new spellings for technical expressions reformers have not yet classified (and have possibly never even heard of), or whether to adopt the *current* attitude of most of the German scientific, technological and industrial world and ignore the reform.

Due to the fact that most specialised literature currently conforms to the old spelling model, and that technical authors of internal as well as external industrial reports will obviously have considerations and priorities far more crucial than changing their own long-established spelling habits, the author felt tempted to opt for the second alternative, especially as any tampering with spelling would adversely affect carefully contrived sorting arrangements. Instead a compromise was adopted. A large section of the book focuses on translation *from* German and adopts twentieth-century spelling conventions. Although certain alternative versions appear in the TPD and TCD, there is little space available for this information owing to structural restrictions. Translators interested in new spellings should consult the Thesaurus instead.

English has been subjected to recent spelling reforms too. Regrettably, these do not make it any easier to distinguish the radically different pronunciations of words like *brought*, *cough*, *drought*, but concentrate on whether to write for instance *realization* as opposed to *realisation*. Nevertheless, if full spelling conformity does ever occur within either the English- or German-speaking worlds it is likely to result not from the activities of spelling reformists but from the free market itself.

3. Text Typology

A good translator tries to capture the message of a source text as accurately as possible and makes finely tuned adjustments to the final version to suit the *text type* (essay, memo, report, etc.) and the intended recipient. The book concentrates on *meaning* rather than *message*, and deals with the *cognitive* aspect of terminology (Sager, 1990), relating *linguistic forms* (terms) to their *referents* (conceptual content). It tends to leave final adjustments to the reader's common sense.

Nevertheless, improvements in this area of language study could soon be taking place. In the modern era of globally interacting data systems there is no reason why lexicographers should not produce well-organised large-scale *collocation dictionaries* along the lines of the TCD. When space restrictions and organisational complexity are not paramount considerations, as in on-line dictionaries, additional translated versions could be included and possibly differentiated according to a set of symbols denoting *text type*. Further research is necessary first, however, in order to establish and classify the various *text-type categories* and reach international agreement on these categories.

4. Terminology Processing

The dictionary arrangements of the book represent practical applications of theories proposed by a long line of linguists, semanticists and terminologists over several generations, ranging from Eugen Wüster to John Lyons to Juan Sager.

Sager (1990) employs various examples from technical language and stipulates the following *types of relation*:

caused by	a counteragent of
a product of	a container for
a property of	a method of
an instrument for	a material for
a quantitative measure of	a place for

He also differentiates *facets* of a generic relationship in an interesting ISO example reproduced briefly (in the style of this book) below:

1 antifriction bearings
differentiated by type of rolling bodies
11 roller bearings
12 ball bearings
by number of rows of rolling bodies along the bearing axis
13 single-row bearings
14 double-row bearings

15 multi-row bearings
by type of forces
16 radical bearings
17 radical axle bearings
18 axle bearings

And Sager lists different *subtypes* of partitive relationship:

1. *atomic constituents* (e.g. the characters of a character set)
2. *groups of parts* (the four suits in a deck of cards)
3. *optional constituents* (the radio of a motor car)

etc.

The e-book side-steps the problem of terminological relationships with different *facets*, by incorporating relevant information either implicitly in the engineering chapters or explicitly via additional thesaurus definitions. Of Sager's list of relational types, one is used directly:

a quantitative measure of

and corresponds to the descriptor *m* (a measurable parameter of). Others appear indirectly via two descriptors *t* (type of), *u* (used in connection with):

a product of t: product; u: …	*a container for* t: container; u: …
a material for t: material; u: …	*an instrument for* t: instrument; u: …

Others can occur directly via the descriptor *d* (designates):

d: a reaction caused by …	d: a property of …
d: a method of …	d: a place for …

Early authors, such as Lyons (1977) or Palmer (1975), look more closely at the *logical* aspects of relations themselves and distinguish between categories such as *symmetrical, transitive, reflexive.* The Thesaurus takes account of this approach but without indicating these aspects directly.

5. Hierarchic Organisation

Another early linguist turned terminologist, along the lines of Wüster/Sager, is Helmut Felber (1980), who quotes the following *engineering* example of what he calls a "genus/species system of concepts", a *generic* hierarchy:

1	(vehicle)
11	land vehicle
12	seacraft
13	aircraft
131	(lighter-than-air aircraft)
1311	balloon
1312	airship
132	(heavier-than-air aircraft)
1321	glider
1322	kite
1323	airplane
1324	rotorcraft
14	spacecraft

Felber also cites a biological example of a system involving *partitive* relations:

1	human body
11	head
111	mouth
112	nose
113	eye
1131	eyeball
1132	eyebrow
12	trunk
13	limbs
131	arm
132	leg

Felber's examples are given in the form of tree diagrams but are quoted accurately above, with the exception of the *bracketed* expressions.

Some writers on linguistics do not make it clear whether their semantic hierarchies depict *terms* or *concepts*. The author would disagree that there *is* an English term *vehicle* covering *land vehicles* and *seacraft*, etc. There seems to be a *concept* but no *term*, a *lexical gap* in English. Hence the entry above is bracketed.

Similarly, the expressions *lighter/heavier-than-air aircraft* are more likely to denote *concepts* rather than *terms*, as engineering does not usually tolerate clumsy expressions like this, any more than natural language does. This is not to say that Felber is wrong. He quotes both a BS (British Standards) definition of *rotorcraft* in terms of *heavier-than-air aircraft*, and a separate BS definition of *heavier-than-air aircraft* itself. But in such cases, if the term is truly used in engineering, an abbreviated form such as *HTA aircraft* is more likely to emerge.

Another slight oversight in Felber's example concerns the terms *seacraft, aircraft, spacecraft* themselves. These belong to a special class of nouns whose plural is invariable (e.g. *one aircraft, fifty aircraft*) and this valuable information (especially for non-native English speakers) is absent. Moreover *seacraft* may belong to a slightly different category to *aircraft/spacecraft*. A theoretical singular form may exist, but it is barely used if ever. Thus the first two examples below represent normal English statements. But not the third.

The Boeing B52 is an aircraft.
Gemini 1 was an early spacecraft.
(*)The QE2 is a seacraft.

Felber's second hierarchy has no lexical gaps but it seems to imply that *eyebrow* is part of the *eye* in the same way that *eyeball* is. Moreover, he realises the problem of non-singular entities, and includes the conception *limbs* rather than *limb*, but does not extend it to *arm/leg*. Similar examples of structural dilemma occur in engineering, where the concept *atom* involves the singular constituent *nucleus* but a specific number of *electrons*.

Nevertheless, apart from the slight inconsistencies, Felber's examples reveal that *hierarchic organisation* is a powerful lexicographical tool for pinpointing both the *denotations* and *connotations* of specialised expressions, as well as highlighting *lexical gaps*. The thesaurus approach devised in the book neatly sidesteps the semantic problems of hierarchic organisation, and does provide information on noun classes, but unfortunately it has difficulty representing lexical gaps. If an English *term* does not exist for a concept there can be no entry in a bilingual thesaurus, regardless of whether a term exists in German

However, there are practical methods of resolving such problems:

i. The TPD employs *brackets* to indicate *descriptive meaning* as opposed to direct *translational equivalence* for entries where inconsistencies occur, e.g. *Reißzeug* (drawing instruments), *Atomhülle* (set of electron shells).
ii. In those Thesaurus entries where a technical term exists in English but there is no true equivalent in German, e.g. the chemical expression *fluid*

that covers both *liquids* and *gases*, the lexical gap is indicated by the symbol *(–)*.

iii. The German Index provides indirect access to *explanations* of German terms for which there are lexical gaps in English and for which concise descriptive explanations in glossary form would be difficult or impossible, e.g. *Atomrumpf* (Chap. 3), *Lastarm, Kraftarm* (Chap. 13), *Höhenschnittpunkt, Kathete* (Chap. 16).

6. Contiguity

An alternative approach to the illustration of *related meaning*, avoiding tree diagrams or hierarchic organisation altogether, is that of Nida (also Lyons, etc.), which involves two-dimensional shapes representing *constituents* of meaning within a *semantic space*, rather like the Set Theory of Mathematics (Ge. *Mengenlehre*). Sets can cover the same area (synonymy), enclose one another completely (hyponymy) or partially overlap (related synonyms, complements, opposites, etc.). Nida differentiates the concept *opposite*, and defines expressions like *polar opposite* (good, bad), *reversive* (tie, untie), *reciprocal* (loan, borrow). He also makes it clear that the related meanings of words like *run, walk, hop, skip, jump* are much closer in semantic space than are different meanings of *run*:

The engine is *running*.
She has a *run* in her stocking.
He is *running* for the office of President.

The method is a good one for revealing *contiguous* relationships of the type *hop, skip, dance*, but not a practical large-scale approach for terminologists or lexicologists. Moreover, like Sager's *facets* of generic relationships, Nida's various categories are representable *indirectly* in the Thesaurus via descriptors like *s* (synonym), *ct* (contrasted with), *cv* (covers), *d* (designates).

Generally speaking, the dictionary symbols adopted by the book involve *simplifications* rather than direct copies of those relations proposed by established writers on semantics. The simplifications seem justifiable, but only time will tell.

7. Speech Acts

The ideas of certain linguists and linguistic philosphers (Searle, Austin et al.) on so-called *speech acts* may seem irrelevant to technical translators at first sight, but this is not the case. Though natural language has secondary functions (*social, aesthetic, ritual,* etc.), it is assumed that the main function of both natural and specialised language is *communicative.* This is true, but large sections of mathematical and scientific text, as well as material concerning engineering design, employ a language function closer to the category *ideational*: involving the formulation of ideas.

The problem for translators is that sometimes lexical gaps or different knowledge structures exist in fundamental areas *crucial* to the formulation of ideas. Even general language contains nuances difficult for translators to work with. Wagener (1978) quotes the German expression *Schaum,* which has two very distinct interpretations in French *écume, mousse* and three in English: *foam, froth, lather.* But the difficulties involving Wagener's *Schaum* are minor compared to the technical translator's quest to find English equivalents for *Elektronenhülle* or *Atomrumpf* (Chap. 3) or decisions as to whether to substitute *resistance* or *impedance* for the German *Widerstand* (Chap. 2). Before terminology can be standardised and stored efficiently in multilingual term banks it is necessary to standardise concepts themselves. In the world of engineering this is by no means easy.

8. Error Analysis

The brief criteria proposed for the evaluation of errors by technical translation assessors (Lex.7) may irritate certain linguists, like Pym, Kussmaul, Hönig, who dislike approaches based on a system which they regard as *binary*: either *right* or *wrong* with no stage in between. Kussmaul (1995) employs an interesting example where the direct translation of *trailer* into German as *Sattelschlepper* is declined in favour of *Lastwagen,* a term normally translated by *lorry* or *truck.* He also quotes examples of *lexical gaps* in language and illustrates difficulties involved in translating the terms *haricot* and *kidney bean* into German, or the British legal terms *barrister, solicitor.* Kiraly (1995) cites other authors (Röhl, Michea), in stating that translation does not imply "transposition from one language to another" but "playing two different keyboards". His case studies for *translation pedagogy* try to demolish the persistent student image of their language instructors as "guardians of translatory truth".

The book does not contest the theories of other authors on error analysis. In fact it produces a variety of parallel examples which might substantiate them, especially among fundamental engineering conceptions such as *conductance*, *impedance*, *tension*. But instead of including lengthy discussions of the *gravity* of particular errors, the book concentrates its attention on providing sufficient background information to enable translators to *avoid* them. If translation involves "playing two keyboards", technical translation uses several more.

9. Sememe, Chereme

If three mathematicians from opposite corners of the world but with similar interests were stranded, without their interpreters, in a hotel room they would soon start communicating with one another, discussing mathematics. An Iranian businessman, a Japanese clerk and a Polish railway worker might, under similar circumstances, discover a common interest in *football*. They too would hold a kind of conversation, one consisting largely of the names of *players*, *teams* and important *matches*. Wagener quotes examples documented by earlier linguists (e.g. William Stokoe) of American Indian and African Bushmen languages which are based not on semantic units of mutual recognition, so-called *sememes*, but on sign units or *cheremes*. Some palaeontologists are of the opinion that neanderthal man employed a similar (though more advanced) means of communication, in conjunction with a smaller set of speech sounds than modern man. And there are methods of communication analogous to a language based on cheremes that exist in various areas of science and technology.

Mathematicians communicate via an internationally recognisable system of *symbols* rather than body signs or hand signals. Chemists, atomic physicists, electronic circuit designers and other scientists and engineers employ their own respective, mutually intelligible symbols. Technical texts may well contain symbols which themselves carry meaning and can assist translators to locate correct L2 equivalents for difficult terminology. The book introduces linguists to concepts of the type *physical quantity*, *dimension*, *unit* (Chap. 1) which are fundamental to all areas of engineering and provide vital clues to the translation of German polysemes like *Spannung*, *Widerstand*, *Leitfähigkeit*. It also presents chemical, electronic, mathematical and many other symbols. But this is just the tip of the iceberg. More research needs to be done in this area so that translators can profit from this additional *extralingual* facility.

10. Standardisation of Nomenclature

Possibly, in the wake of repercussions from the famous confrontations of the fifties between *behaviourists* and *mentalists*, the Skinner versus Chomsky disciples, Lyons, Nida, Palmer and other semanticists of the seventies epoch were influenced by logical or algebraic formalisation schemes, such as *propositional* or *predicate calculus*. But Wüster (1974), towards the end of his long career, was already applying this knowledge to the more practical aspects of terminology and lexicography. Wüster's pioneering work dating from as early as 1931 led to the formation of committees for the *standardisation of nomenclature*, one of the most influential being the ISO (International Standards Organisation). His epic book *The Machine Tool* marked the birth of a new type of glossary, and possibly of a new discipline: *terminology* itself. Though some authors were a little unhappy with the designation of the new field and introduced expressions like *terminography* (Sager), *terminological lexicography* (Felber) to cover specific aspects of what seemed to be a new profession, *terminology* is now an established component of modern university curricula relating to language study.

In response to the need for standardisation, a number of *term banks* arose over the years, such as *Eurodicautom* compiled for the Commission of the European Community. Some linguists of the eighties were disturbed by Britain's initial reluctance to contribute to this work, and there were even suggestions that, without Britain's involvement, terms or usages might become established which would be unacceptable or even incomprehensible in Britain itself. But linguists, or rather terminologists, tend to underestimate *market forces*. Electronics enthusiasts had no trouble discarding the antiquated terminology from the days of valve circuitry, *resistance, condenser, coil* in favour of *resistor, capacitor, inductor*, nor with the transition from the long-established unit of frequency *cps* (cycles per second) to the recommended SI unit *hertz*. Similarly, chemists have swiftly adapted to the more consistent, new designations of chemical compounds, *dinitrogen monoxide*, etc., quickly discarding the long-established forms like *nitric oxide, nitrous oxide*, and computer enthusiasts in Britain differentiate between *TV programme/computer program, brake disc/hard disk* almost subconsciously.

11. Translation Approaches

Twenty-five years ago, courses on technical translation were given by eminent linguists and part-time lexicographers owning vast libraries of technical dictionaries and large filing cabinets of card indexes. These assessors did indeed regard themselves in the words of Kiraly as "guardians of translatory truth". Confidence in the old masters ebbed when universities began introducing *subsidiary courses* (Ge. *Ergänzungsfach*) on Economics, Law and Engineering, which helped students to understand what they were translating and encouraged them to consult *literature* rather than antiquated alphabetic dictionaries. Translators with access to libraries at universities where these subjects were also taught as main subjects had a substantial advantage over those taught at smaller institutes. But free access to modern Internet facilities has now evened out this gap, and a new *on-line approach* to translation is emerging.

The author has avoided any comment on these approaches so far, as they have not yet crystallised and are bound to change with each advance in software. But a tentative recommendation can be given to future translators. Now that first-hand access to specialised literature is finally available it should be properly analysed, and not just skipped through on the monitor. The book provides the necessary background knowledge to begin a career in technical translation, but a *systematic* approach to data analysis and storage is necessary if individual translators wish to improve upon this basis.

12. Future Technology

If a book of this type had appeared in 1970 it might have contained wild predictions about future technology in the year 2000, envisaging permanent space stations on the moon and trips by astronauts to Mars. But none of this has come about.

The industrial basis of engineering has expanded, but it has not changed beyond recognition. It is professions themselves that have changed dramatically. A motorist having occasional problems with his *starter* during the fifties would consult a garage mechanic who would locate the fault (possibly a loose connection) and eliminate it. Nowadays most mechanics order a new starter regardless of the fault, and would probably be unable to fix it anyway. Neither the technology of starter motors nor their terminology has changed much over the years, but increased specialisation, indeed monopolistic over-specialisation has not necessarily led to a reduction in the frequency of translation assignments for such areas. Faults in engineering components still have to be described and analysed, even when the component itself is eventually discarded. In fact, demands on *translation quality* are more important than ever, now that customers have access to translation memory and alternative sources of cost-free machine translation.

13. Technical Language

The book began with the commitment to provide interested learners with a working knowledge of *technical language.* Some readers may have been sceptical that there is such a thing as "technical language", expecting the book to present simply a neat collection of material on various unrelated engineering disciplines. But all engineering branches are rooted in natural science and, just as linguists share a basic understanding of areas as diverse as *phonology*, *morphology*, *syntactics* and *semantics*, all engineers share a common basic language, which develops in different ways according to the industrial environment and the technological expertise available in the country concerned. The engineering chapters demonstrate that certain expressions recur throughout technology and many of these are *polysemous.* Moreover, technical language sometimes involves different prepositional or verbal constructions to those anticipated from general language, and even slightly different grammatical rules.

School-leavers embarking on foreign-language courses hope that one day their competence will improve to such an extent that they are indistinguishable from native-speakers. But improvements in didactic methods of language instruction have recently brought this seemingly unattainable goal closer than ever before. Likewise, this book removes the *drudgery* and the *guesswork* which has dominated technical translation for so long. It brings a working command of technical language within the grasp of any proficient linguist, thus providing a solid basis for the acquisition of professional translation skills.

14. Reference

The author belongs to the generation of translators that followed the ancient guardians of translatory truth, and has acquired a substantial library, not of technical dictionaries but of first-hand scientific and engineering literature, in both British and American English and in German. The reader might expect the book to close with a long ragged list of these references. But there is no real need for this, now that on-line facilities provide instant access to similar material and can suggest better, up-to-the-minute reference sources.

The book encourages translators to study *semantics* and linguists to study *technical language.* The semantic approaches of Lyons, Nida, Felber and Sager provided valuable inspiration for the Technical Thesaurus, and other linguists are quoted whose ideas are reflected in the book too, some of which go back a long way. Just as the *music* of Gershwin, Lennon/McCartney and Carlos Santana has withstood the test of time, so the ideas of certain early contributors to the fields of *linguistics, semantics* and *terminology* are still valid today. This Appendix admittedly picks the raisins out of the cake but may encourage some members of the modern teaching elite to look at these areas for inspiration once again.

Finally, attention is drawn to particular statements made by two influential founding fathers of Linguistics:

i. *The meaning of a word is its use.* (Wittgenstein)
ii. *You shall know the meaning of a word by the company it keeps.* (Firth)

Though quoted out of context, Wittgenstein's statement could justify the *collocational* approach to terminology organisation illustrated by the book, and Firth's statement the *thesaurus* approach.

Appendix 2

American-English Survey

The author specialises in British English translation, which is reflected in the book. Fortunately, there are few areas of technology where serious discrepancies exist between British and American terminology. But, where they do occur, they need very close attention. Every effort was therefore made to include American alternatives at each stage of the project. Key sections, especially of *Chapter 8 — Automotive Engineering* and of the disk *Appendix* were carefully checked against the existing literature, and tables were modified with the advice of available American native speakers. Despite this, at almost the final stage, the book received a surprising but well-meaning attack on the very sections and the very topic for which so much trouble had been taken.

Apart from occasional terminological discrepancies themselves, one real problem was that all terminology, including American terminology, was discussed in British English. For American speakers, especially non-native speakers, this could be misleading. The glossaries enable translators from German to distinguish British terminology that has no place in American, and vice versa, but though examples of where British English accepts American were specific, the converse was not true. Simple last-minute alteration of the disk would have provided a mish-mash unacceptable to either British or American translation enthusiasts, and involved the loss or obscuring of valuable examples quoted elsewhere in the text sections. Instead, the "corrections" stipulated by the final critic are presented in the form of neat tables below.

1. Transatlantic Glossaries

The first table concerns terminology from the *Transatlantic Lexicon* (disk appendix, Figure A1) that, by virtue of the British-orientated dictionary structure is occasionally misleading for translators into American. The information is arranged in order of German, and the first three columns are exactly as on the disk. It is the final column *Am(2)* that contains the vital supplementary information for non-native American speakers. Terms are either qualified, e.g. *precinct (u: police area)*, or related to terminology of the third column via two descriptors:

< *tramp, vagabond* — implying some Americans use these terms too
> *renter*, > *playground* — some Americans use these instead

Supplement to Figure A1.

German	British	Am (1)	Am (2)
Bezirk	district	precinct	precinct (u: voting)
Bezirk	district	precinct	precinct (u: police area)
Bezirk	district	precinct	district (u: school area)
Brieftasche	wallet	billfold	< wallet
Eimer	bucket	pail	< bucket
Geschäftsinhaber	owner (of a business)	proprietor	< owner
Herbst	autumn	fall	< autumn
Hose	trousers	pants; dress slacks	< trousers
Krug	jug	pitcher	< jug
Landstreicher	tramp	hobo	< tramp, vagrant
Notizblock	note pad	scratch pad	< note pad
Rechtsanwalt	lawyer	attorney	< lawyer
Schnürsenkel	shoelace	shoestring	< shoelace
Schulhof	playground	schoolyard	> playground
Strumpfhose	tights	pantie hose	< tights
Untermieter	lodger	roomer	> renter
Wasserhahn	tap	faucet	< tap
Wohnwagen	caravan	trailer	< mobile home

A supplement was also requested for the *Transatlantic Thesaurus* of Figure A2.

Supplement to Figure A2.

Term	Implication, Br.Eng.	Implication, Am(1)	Implication, Am(2)
bill	ex: hotel bill;	ex: 10-dollar bill;	< ex: hotel bill;
chips	u: meal; ex: fish & chips;	u: popcorn, candy, etc.;	> d: thin sliced crispy potatoes
purse	a: loose change	t: handbag	< d: change purse
subway	t: pedestrian walkway	t: underground railway	> t: underground railroad
tack	u: carpeting, panelling	u: office stationary	> u: office products
trailer	u: load being towed	t: mobile home	< u: load being towed
wallet	u: banknotes; a: pocket	u: cash; a: bag;	> u: cash
wash up	u: dishes, plates	u: people	< u: people, dishes

2. Automotive Terms

The same descriptor devices as above are employed to indicate supplementary information to Figure 8A (Chapter 8). Thus *>inner tube*, *>choke*, *>antifreeze* mean that some American customers, terminologists, translator educators, etc. dispute the terms quoted in column 3, i.e. *air tube*, *air strangler*, *defreezer*, or merely prefer those of column 4. Likewise *<accelerator*, *<gas*, *<rotary* indicate alternatives.

Other microglossaries of the same chapter do not contain different American versions. But in two cases, their inclusion was subsequently deemed necessary:

Figure 8D: Brake Assembly, Hydraulic System
Figure 8E: Steering, Suspension, Body, Windscreen

Supplement to Figure 8A.

German	British	Am (1)	Am (2)
Autobahn f	motorway	freeway	< interstate, interstate highway
Benzin n	petrol	gasoline	< gas
Blinker	indicator	turn signal	< blinker
Bremslicht n	brake light	stop light	> brake light
Dachgepäckträger m	roof rack	car-top carrier	< roof rack
Fahrpedal n	accelerator pedal	gas pedal	< accelerator
Frostschutz m	antifreeze	defreezer	> antifreeze
Kreisverkehr m	roundabout	traffic circle	< roundabout, rotary
Limousine f	saloon car	sedan car	> sedan
Motorrad n	motorbike, motorcycle	motorbike	> motorcycle, ex: BMW, Harley Davidson
Nebenstraße f	side road	branch road	> side road
Reifenschlauch m	inner tube	air tube	> inner tube
Rückstrahler m	reflector	bull's eye	> reflector
Rückwärtsgang m	reverse gear	back-up gear	> reverse gear
Vergaserluftklappe f	choke, choke disc	air strangler, choke	> choke
Verkehrsinsel f	traffic island	channelizing island	> traffic island
Wechselstromlichtmaschine	alternator	ac generator	< alternator
Wegweiser m	signpost	direction post	> sign post, signage

Supplement to Figure 8D and E.

German	Figure	British	Am(2)
Bremsleitungen pl	Figure 8D	fluid lines	brake lines
Bremsseil	Figure 8D	handbrake cable	brake cable
Bremsträgerplatte	Figure 8D	backplate	back plate
Chromteile	Figure 8E	chromework	chrome
Handbremshebel	Figure 8D	lever	handbrake lever
Karosserie	Figure 8E	bodywork	body
Lackierung	Figure 8E	paintwork	paint
Parkbremse	Figure 8D	handbrake assembly	handbrake
Radstand	Figure 8E	wheel spacing	wheelbase
Türverkleidung	Figure 8E	door cladding	door cladding
Türverkleidung	Figure 8E	door cladding	door lining (co: soft fabric)

3. TPD Entries

The knowledgeable critic responsible for the supplements above also kindly analysed the TPD and Thesaurus for discrepancies. The result appears below:

German	Field	British	Am(2)
außer Betrieb	GEN	non-operational	out of service, down
Automatik	AUTO	automatic gears	automatic transmission
Bahnschranke	RAIL	level-crossing barrier	grade crossing gate
Bremsspur	AUTO	tyre tracks	brake marks, skid marks
Elastizitätsmodul	PHYS	elasticity modulus	modulus of elasticity
Faßhahn	HOUS	spigot	spigot, tap
Gabelschlüßel	TOOL	open-jaw spanner	open-end wrench
Gummiband	GEN	elastic band	rubber band
Hoch	METR	anticyclone	high pressure area
Kupplungsbelag	AUTO	clutch lining	clutch facing
Leitungsnetz	ELEC	distribution network	distribution grid
Netzanschluß	ELEC	mains socket	electrical socket, electrical outlet
Normalbenzin	AUTO	3-star petrol	regular gas
offener Güterwaggon	RAIL	goods truck	open freight car
Ottomotor	AUTO	petrol engine	gasoline engine
Prüfstand	MECH	test-bed, test block	test rig, test stand
Radstand	AUTO	wheel spacing	wheelbase
Radträger	AUTO	wheelbase	steering knuckle
rechter Winkel	GEOM	rightangle	right angle
Sechskantstiftschlüßel	TOOL	Allen key	Allen wrench
Spülbecken	u: kitchen	sink	kitchen sink
Spülbecken	a: toilet	lavatory pan	sink, lavatory
Stromnetz	ELEC	power network	power grid
Superbenzin	AUTO	4-star petrol	premium gas
Torsionsmodul	PHYS	rigidity modulus	modulus of rigidity
Versorgungsnetz	POW	supply network	supply grid
Winkel	GEOM	setsquare	square
Zündkerzenschlüssel	IGN	plug spanner	spark-plug wrench

It can be frustrating for translator trainers when students repeatedly come up with the same inappropriate terminology substitutions, obtained from a valuable source not specifically designed for their purpose. The irritation sensed by the critic concerned, on observing a number of L2 substitutions presented in isolation that were inappropriate to his language variant, is therefore understandable. Familiarity with the table extensions above, however, should enable translators into American English to avoid such pitfalls. The additional substitutions are readily accessible (red eye) anywhere on the disk in the form of three compact microglossaries: Automotive Engineering (Figures 8A–E), General Engineering (TPD/TT), General Language (Figures A1–2).

A few other discrepancies exist, for instance *aluminium (Br.)*, *aluminum (Am.)* or the many derivatives of *sulphur (Br.)* in the terminology of chemical compounds, e.g. *sulphuric, sulphurous, sulphate, sulphite, sulphide,* that require the mental substitutions *sulfur (Am.)* — *sulfuric, sulfurous,* etc. But most of these are mentioned in the chapters or appear in the microthesauri.

Appendix 3

Overall Survey

The previous unit re-examines the book from a specific perspective, that of a German-speaking technical translator into American English who might be misled by the British terminology and definitions. This unit surveys it from another perspective, that of *future analysts*. It points out a variety of possible misconceptions, justifies approaches taken in the light of the way the book was written, and discusses the reader's realistic expectations of the work itself.

1. Disk Format

Technical translation is an enormous field and different readers encountering this book will inevitably have different preferences and expectations. Some will expect to insert the disk, type in a term and immediately discover scores of occurences in completely random contexts or simple database structures, an approach involving *single-click access* without *user intelligence*. The disk provides at best *double-click access* with *minimal user intelligence*:

Load disk
Click 1 eye button (blue, green, brown, purple)
Click 2 page (GE-B, TPD-F, 3.4.1, Figure 8B, etc.)

Thus Volume 3 (GE, EG) displays immediate translations of technical terminology. Another click at the appropriate eye button enables immediate access to information concerning this terminology. Any *fumbling* (clicking on the wrong eye button or wrong section) costs time, though this and other less obvious potential difficulties are easily avoidable if the disk instructions are studied and properly implemented (contents list, multiple windows, etc.).

The disk is formatted in HTML (HyperText Markup Language), a facility used recently to format the Web itself. It could be argued that with other electronic information management environments with a *global search engine, terminology management, database output facilities*, etc., such as XML or XSL, the limitations of the HTML disk format mentioned above need not occur. But it should be borne in mind that the project began, not as a database, but as conventional book that outgrew any conventional size. The decision to employ HTML was taken at a time when no assistance was offered by either the publisher or the author's academic institute, and when the author was grateful to find *any* software enthusiast prepared to spend so many months patiently reformatting a vast collection of Word files into electronic form. If the decision to use HTML was wrong, so be it. Without it the book would not have appeared at all.

2. High-Tech, Low-Tech

Like technology, the ideas of linguists, translators and translator educators are perpetually changing. This is reflected in the range of university curricula offered. Some institutes teach language and convey mainly general translation skills, but provide technical translation as a valuable option. Others, at the opposite end of the scale, teach specialised translation disciplines — Engineering, Medicine, Law, Business — virtually from the first semester onwards. Some provide joint honours curricula, combining language teaching with a specialised discipline itself. Most institutes now provide facilities for database management, translation memory, Web searching, etc. and try to make the transition from *low-tech* to *high-tech* translation, for graduates entering full-time employment, as smooth as possible.

But without *low-tech* there could be no *high-tech*. The principles underlying the loudspeaker system of a football stadium or the minute circuitry of a hearing aid, the starter of a heavy goods vehicle or the motor of an electric toothbrush, the products of a nuclear reaction and the substances employed in medical diagnostics contain many basic parallel similarities. *Low-tech* terminology tends to remain fairly stable from one decade to the next but, because it often precedes the existence of the Web, it is still notoriously difficult to specify accurately.

The book provides low-tech terminology and a low-tech approach to scientific and engineering translation. But there is no shame in this. It enables the linguist to become an intelligent, respected, equal partner of the technologist rather than a dogsbody who perpetually confuses *power* with *performance*, *current* with *voltage*, *resistivity* with *resistance*, and makes scores of similar elementary blunders for which the translation customer can see neither rhyme nor reason. In short, the book fulfils the objectives it began with: the description and specification of simple *technical language*, that elusive dialect of communication understood at least passively by all scientists, engineers, mathematicians and technologists regardless of their specialist discipline or field of interest.

3. Popular Misconceptions

The chapter on *Automotive Engineering* contains plenty of terminology, a fair number of useful coloured illustrations, and raises some interesting semantic points: *ignition coils* are not *coils*, *water pumps* do not pump *water*, *high-tension leads* have nothing to do with *tension*. It paves the way for the passive stimulation of similar intellectual processes in less familiar fields: a *saw-tooth generator* is not a *generator*, a *multivibrator* does not *vibrate*, the *base* of a *transistor* has nothing whatever to do with the *bottom*. Terminology discussions invite the reader to step back for a moment to contemplate the terms concerned, thus improving familiarity with them and enhancing subsequent correct usage in practical translation situations. One critic felt that the book seems to support a prescriptive approach to translation, trying to impose terms like *ignition transformer*, *coolant pump*, *HV-lead* on industries where they are simply not used. A closer look at any chapter reveals that nothing could be further from the truth.

A reader also complained about the alleged *"brevity of the microthesauri"*. Another wanted all the *microglossaries* to appear in one place. The disk chapters concentrate on the *difficulties* of translation, not on the identification, collection and registration of *all terminology* relating to the field concerned. The prime purpose of the microglossaries and microthesauri is to provide sufficient entries to indicate the essential *terminological* and *conceptual relationships* relevant to the chapter. The disk is neither a *database* nor a conventional *e-book*, neither a *lexicon* nor a *text book*, neither a simple *translation tool* nor purely an *instructive aid*. It is all of these things, and yet none of them. *It is what it is*. A similar labelling difficulty applies to the conception *microglossary*.

4. Coverage, Detail

Many translator educators have two main areas of specialisation, *Automobile* and *Computer Engineering*. The book does not provide a great deal of *detailed* information on these two fields as this is available abundantly and covered better (with more explicit diagrams, etc.) by other electronic media. Instead it uses these familiar fields to extract linguistic models for *general* discussions of technical translation complexities, that are applied elsewhere in the book. Emphasis on detail is more evident, nonetheless, in fields that tend to be overlooked in a classical translator education programme (the other 14 chapters) and a balance is struck among the fields considered. Thus the description of areas like *Semiconductors, Nucleonics, Electronics, Machine Technology, Heavy Electrical Engineering, Chemical Engineering* tends to be more explicit than the two relatively straightforward main areas which most translator educators seem to want to examine in isolation first.

These neglected fields in conventional schemes for translator training have important features of their own. Certain key concepts are shared, for instance by Chemical, Nuclear and Materials Engineering, Mechanical, Aeronautical and Construction Engineering, Electrical and Electronic Engineering, and it is interesting for the reader to see how terminologies of German can widely diverge from those of English in this respect: *Flüssigkeit* (liquid, fluid), *Auftrieb* (lift, buoyancy, upward thrust), *Kondensator* (condenser, capacitor). Having overcome the conceptual hurdles, the reader will find that translation itself is not necessarily more difficult in these fields. Indeed, the insights and perspectives obtained by *students* who translate material from a broad range of engineering disciplines open up new job markets to them and stand them in good stead when later the only way to supplement this expertise is by conventional electronic search methods.

Another feature of the disk is that it encourages translators to consider *different* lexicological approaches that might be applicable to their own data management schemes. Figure 10A illustrates the advantage of reverse-sorting arrangements in connection with the names of chemical substances: *calcium carbide, zinc iodide, hydrogen sulphide.* Figure 9B contains a glossary devoted entirely to verbs: *bore, braze, cast, forge, grind,* etc. The mathematical glossaries of Figures 16A–G all employ simple hierarchic organisation with great success. And most chapters contain *microthesauri* inter-relating the basic terms of the fields concerned and highlighting useful linguistic aspects: *polysemy, homonymy, contrast, usage, association.* There is no time to consider more creative approaches to terminology management when students are forced to rely completely on conventional electronic search methods throughout their education. But, without them, life as a translator can be very dull.

5. Possible Omissions

The author's approach to technical translator training is to give students as broad a base as possible, so that when as professional translators they are subsequently obliged to specialise they have a solid background enabling them to switch fields at any time, just as a qualified mechanical engineer might later specialise in aircraft design, or a chemical engineer might seek employment at an electronics company. Many academic institutions currently take the opposite approach and oblige their students to stick rigidly to just one or two disciplines. By using the multifunctional disk for a single translation purpose, they will inevitably discover omissions, terminological and otherwise, sooner or later. The book should therefore be viewed in terms of the *quality* of the information provided and not the *quantity*.

More diagrams, more glossaries, more definitions, indeed more of everything would be nice. But the disk already constitutes a vast manuscript, and *more of everything* does not necessarily imply that the additional material could be properly or usefully absorbed by the reader. The disk contains information that should be persistently at the fingertips of every proficient technical translator, and a little bit extra into the bargain. It is the work of one person and is absorbable as such — even by student translators, in appropriate doses, with helpful guidance from their peers (Kiraly 2000).

6. Materials, Presentation

In the words of one critic, the disk is "*a paper-orientated presentation of materials stored on an electronic medium*". Though possibly intended as a slighting remark, this statement is nevertheless not far from the truth. The book provides insights and materials from the author's own experience in freelance translation, but is adapted more to his current activity of *translator training* than to the total requirements of a highly specialised translator working daily in any one of these fields. Hence the materials presented are not so much what is usually taught at universities, as what should be taught, in small packages at least, so that the next generation of translators will not have to flounder for hours trying to produce meaningful target renderings of *basic* terminology, from information scattered all over the Web, and need no longer automatically inherit the mistakes of their predecessors. But to get the best out of the disk, it needs to be read occasionally in small stages in consecutive order, just like the handbook.

Some readers might find the electronic presentation itself substandard compared to that of other *e-books* designed with database assistance from the outset. If so, they are welcome to their opinion. The material itself is not, and anyway sooner or later all translators are glad to stop clicking for a while to read something properly and systematically. Obviously the disk cannot provide immediate answers to everything the translator wishes to know. Nor does it necessarily provide rapid answers. It supplies translators with the means to improve their background knowledge, so that many questions no longer need to be asked.

Index